Advances in Passive Microwave Remote Sensing of Oceans

Advances in Passive Microwave Remote Sensing of Oceans

Victor Raizer

CRC Press
Taylor & Francis Group
Boca Raton London New York

CRC Press is an imprint of the
Taylor & Francis Group, an **informa** business

CRC Press
Taylor & Francis Group
6000 Broken Sound Parkway NW, Suite 300
Boca Raton, FL 33487-2742

First issued in paperback 2019

ISBN-13: 978-0-4987-6776-7 (hbk)
ISBN-13: 978-0-367-87816-0 (pbk)

Library of Congress Cataloging-in-Publication Data

Names: Raizer, Victor Yu, author.
Title: Advances in passive microwave remote sensing of oceans / Victor Raizer.
Description: Boca Raton : Taylor & Francis, a CRC title, part of the Taylor & Francis
   imprint, a member of the Taylor & Francis Group, the academic division of T&F
   Informa, plc, [2017] | Includes bibliographical references.
Identifiers: LCCN 2016043405| ISBN 9781498767767 (hardback : acid-free paper)
   | ISBN 9781498767774 (ebook).
Subjects: LCSH: Oceanography--Remote sensing. | Microwave remote sensing.
Classification: LCC GC10.4.R4 R38 2017 | DDC 551.460028--dc23.
LC record available at https://lccn.loc.gov/2016043405

Visit the Taylor & Francis Web site at
http://www.taylorandfrancis.com

and the CRC Press Web site at
http://www.crcpress.com

Contents

List of Figures

List of Tables

Preface

This book is about passive microwave diagnostics of the ocean environment. The aim is to demonstrate the capabilities of passive microwave techniques for enhanced observations of ocean features, including detection of (sub)surface events and/or disturbances. This book outlines the benefits and limitations of these methods and also establishes and maintains a better knowledge of physical principles of passive microwave diagnostics. It is an important milestone for successful and correct geophysical application of ocean microwave data. It is also important for advanced developments.

Between 1980 and 2010, dramatic progress in microwave technology and computer science was achieved, allowing specialists and scientists to provide systematic remote sensing observations of Earth's environment. Today, there are a number of microwave radiometric systems designed by many countries and under the leadership of international communities from the United States, Canada, Europe, Japan, South Korea, China, India, and Russia. These remote sensing systems operate at millimeter and centimeter ranges of electromagnetic wavelengths providing regular space-based observations of atmospheric parameters, precipitation, and ocean–atmosphere interactions, including tracking and prediction of hazard events (storm surges and hurricanes).

Owing to objective limitations in the satellite technology industry and operational cost, most space-based passive microwave radiometers/imagers have low pixel resolution, which is approximately 30–100 km depending on microwave frequency and orbital parameters. It is assumed by many people that such a spatial resolution is enough for global seasonal change mapping of Earth's geophysical parameters as well as for meteorological and climate purposes. But it does not seem to be enough for the registration of ocean dynamic features and localized events.

Indeed, current passive microwave radiometer missions are not capable of conducting valuable observations of complex nonlinear wave processes occurring in the ocean, although some possibilities for the technical modernization or renovation of microwave remote sensing instrumentation exist and might be considered and realized in the near future.

Meanwhile, the application of high-resolution passive microwave methods for advanced ocean studies is the technological frontier involving new scientific ideas, theoretical and experimental studies, and great results that may change the situation cardinally. In particular, there is a good chance to reveal a number of localized oceanic phenomena and/or hydrodynamic processes through high-resolution multiband passive microwave imagery and digital enhancement of the registered radiometric signatures.

In fact, sophisticated microwave radiometers with resolution from a hundred meters to a couple of kilometers will be capable of providing relatively low-cost complementary maritime surveillance and operational control over restricted ocean areas, including the detection of (sub)surface events as well. To achieve this goal, however, innovative techniques, proper understanding of the problem, and correct intelligent methods of data analysis and interpretation are required. These issues are very important for many remote sensing applications; they are a key focus and a major subject of our multiyear scientific research, which are partially summarized in this book.

In the book *Passive Microwave Remote Sensing of Oceans* (Wiley, Chichester, UK, 1998, 195p.) by I.V. Cherny and V.Y. Raizer, which has been out-of-print for some time, we performed theoretical–experimental analysis of the main ocean microwave characteristics and observable by radiometer/scatterometer effects. Microwave contributions from surface waves and roughness, breaking waves, foam/spray/bubble disperse media, and impacts of thermohaline variations, oil spills, rain, hazard events, and other factors on ocean emissivity have been explored and explained in detail. These data and results still have fundamental meaning in ocean microwave radiometry, polarimetry, and spectroscopy.

However, for the efficient implementation of high-resolution ocean observations, more accurate techniques and models are required. As a matter of fact, we deal with a highly *dynamical, stochastic, multiscale, noisy,* and as a whole, unpredictable natural object, which is the real-world ocean environment. This circumstance eventually leads to a great variability and complexity of the collected microwave remotely sensed data, even to the uncertainty of their physics-based interpretation. Therefore, novel studies and efforts are necessary in order to achieve considerable practical progress in the field of ocean remote sensing.

In this book, I continue the analysis of remote sensing methods, models, and techniques. The book focuses on a *high-resolution multiband imaging observation concept.* This advanced approach provides a new level of geophysical information and data acquisition. Experiences show that the measurement of localized hydrodynamic phenomena and/or events is difficult to provide using just one- or two-frequency low-resolution microwave sensor—imaging radiometer or radar. The reason is natural causes such as scaling, variability, and nonstationary. Microwave responses and relevant signatures are usually frequency-band dependable and have space–time characteristics. In other words, the stochastic nature of the ocean surface affects the microwave measurements dissimilarly at different observation conditions, different electromagnetic bands, and different spatial and temporal scales.

Problems of adequate modeling, correct analysis, and interpretation of multivariable data become critical in remote sensing and always represent a challenging task for serious researchers. In this connection, this book has been significantly updated and rewritten, although the structure of the text remains the same as in the previous one. Note that this book does not replace fully the old one. I believe that our novel ideas and materials will encourage the readers to take the next step in the right direction.

This book is divided into seven chapters. Chapter 1 is an introduction to the subject, including a historical survey, basic elements of microwave theory, technological aspects, and methods of data processing and interpretation.

Chapter 2 outlines the main oceanic phenomena and hydrodynamic processes that can be potentially observed and/or detected using multiband passive microwave radiometry and imagery. This chapter provides a primary knowledge in ocean physics and hydrodynamics needed for a better understanding of remote sensing methods.

In Chapter 3, experimental and theoretical data concerning microwave emission of the ocean surface are overviewed. Impacts of surface waves, roughness, turbulence, foam, whitecap, spray, bubbles, and oil pollutions on the ocean emissivity are analyzed in more detail. The microwave models and numerical examples selected and presented in the chapter demonstrate the current status of research in this field.

Chapter 4 establishes novel *composition principles* of microwave remote sensing (and diagnostics) of oceans. This chapter provides a basis for more accurate physics-based modeling and simulations of microwave radiometric data—signals, images, signatures, their properties, and time and space characteristics. New results are generated using stochastic and deterministic multifactor electromagnetic–hydrodynamic models and numerical methods. I believe that a flexible multifactor approach is more adequate and, perhaps, more realistic for purposes of microwave diagnostics and detection of ocean variables.

Chapter 5 provides the essential concept of high-resolution multiband passive microwave observations. This material covers a number of theoretical, methodological, and technical issues. In particular, I present several model and real experimental examples in order to demonstrate the capabilities of high-resolution microwave imagery in ocean studies. One of the greatest concerns and simultaneously significant advances in remote sensing is the assessment of ocean microwave signatures associated with different environmental processes and events.

Chapter 6 focuses on the potential possibilities to observe sophisticated oceanic events using passive microwave techniques. A number of hypothetical (but realistic) microwave scenarios are considered. I decided to include this particular material because it could offer some guidance for conducting future studies and complex experiments.

The summary and several important problems are presented and discussed in Chapter 7. The concluding paragraphs briefly point out the benefits and advantages of passive microwave observations of the ocean.

The scope of this book includes several interdisciplinary topics related to oceanography, hydrodynamics, microwave technology, physics, numerical modeling, digital data processing, and interpretation. This book represents a motivated introduction and complete-at-the-moment informative description of passive microwave remote sensing of the ocean. The problems outlined may provide readers an opportunity to improve their expertise on this particular subject. The bibliography provides an overview of the experimental and theoretical data collected worldwide. References may help many researchers, students, or simply enthusiasts who wish to take the next step and contribute to the development of the subject.

As a whole, statements, recommendations, and some accomplishments presented in this book are useful for many specialists who work in various geophysical and remote sensing fields.

Acknowledgments

I am deeply indebted to many people, specialists, scientists, colleges, managers, and officials for their great support and interest in this sophisticated problem for the past several years, which made it possible for me to obtain new data and materials that are reported in this book.

My special thanks to Albin Gasiewski and Gary Wick for the collection of incredible imaging data and the creation of advanced hardware and software products.

I also very much appreciate the excellent work of CRC Press team and editors, especially Irma Britton with her kind assistance, who helped me bring this book to the market.

Finally, I am grateful to my wife Elena for her constant encouragement and patience.

Victor Raizer
Washington D.C. Metro Area
Fairfax, Virginia
September 2016

Author

Victor Raizer is a senior scientist, physicist, and researcher with over 30 years of experience in the field of electromagnetic wave propagation, radio-physics, hydrophysics, microwave radiometer/radar, and optical techniques, and Earth observation. He graduated from Moscow Institute of Physics and Technology in 1974. He earned his PhD in 1979 and Doctor of Science, higher doctorate degree, in 1996. He worked with Space Research Institute, Russian Academy of Sciences, Moscow (1974–1996), and then at Science and Technology Corporation, USA (1997–2001) and Zel Technologies, Virginia (2001–2016).

Dr. Raizer was involved in various Soviet remote sensing programs, ocean field experiments, and laboratory and theoretical investigations. He was an active co-investigator in the Joint U.S./Russia Internal Wave Remote Sensing Experiment (JUSREX 1992). His research focused on aircraft optical and microwave measurements, physics-based modeling, and data interpretation. In recent years he has provided a broad spectrum of scientific research, including the development of advanced multisensor observation technology, and the modeling, simulation, and prediction of complex remotely sensed data.

Dr. Raizer has published two books and over 50 scientific papers, reports, and presentations in international symposiums and meetings. He has been a member of IEEE since 2002 and a senior member since 2012.

1

Introduction

This chapter provides a sketch of problems and techniques concerning remote sensing of the ocean. Instrument performance and data processing, and possibilities of modeling and interpretation are considered in a methodological manner. A history of ocean exploration from space and other platforms using passive microwave radiometers is briefly outlined. The objectives are formulated as an integrative research program. The bibliography at the end of this chapter provides the reader with additional comprehensive knowledge about each specific topic discussed in this chapter.

1.1 Basic Definition

Remote sensing has been defined in many different ways. One of them is the following: "The science of remote sensing consists of the analysis and interpretation of measurements of electromagnetic radiation (EMR) that is reflected from or emitted by a target and observed or recorded from a vantage point by an observer or instrument that is not in contact with the target" (Mather and Koch 2011). Microwave remote sensing methods "provide a different and unique view that offers new information about Earth's environment that often can be obtained in no other way" (Ulaby and Long 2013).

Both definitions are correct and acceptable in the context of this book but they do not clarify how to get this new information from remote sensing measurements? Here, we come to interdisciplinary scientific research program, which is usually divided into several independent topics, but it is considered sometimes separate from the main problem, depending on individual experiences and skills.

In fact, unlike other environments, oceans that cover more than two-thirds of Earth's surface represent the most complicated geophysical object for exploration. A majority of global dynamic processes at the ocean–atmosphere interface is difficult to control using traditional *in situ* methods. Among these are multiscale wind-generated waves, stormy situations, frontal zones and currents, turbulent flows, thermohaline (temperature–salinity) circulations, and some synoptic events. These and other large-scale processes occurring at the marine–atmospheric boundary layer can be discovered from satellites using different remote sensing instruments and techniques, including passive microwave radiometry. However, remote sensing and detection of the *ocean dynamic features* demands the application of an integrated scientific approach, including the understanding and comprehensive knowledge of both hydrodynamic and electromagnetic aspects of the problem.

1.2 Instrument Performance

There are two categories of microwave remote sensing instruments: active (radar, scatterometer, interferometer, altimeter, and global position system [GPS] tracker) and passive (radiometer, sounder, spectrometer). The active sensor transmits electromagnetic waves at a certain frequency and then measures the scattering or reflected signal from the investigated object, medium, or surface. In this case, we usually obtain selective information about spatially statistical (geometrical) characteristics of the object.

The passive sensor does not transmit any electromagnetic signal but it measures the thermal radiation emitted by the medium, object, or body itself at the selected (and fixed) microwave band. As a result, we obtain integrated information about the thermodynamic, structural, and physical properties of the medium plus its surrounding environment.

Radar methods are well developed; they are widely used in satellite oceanography for observations of ocean surface features, internal waves, ship wakes, oil pollutions, boundary layer convective mixing processes, mapping of ocean floor and sea level (topography), and Arctic and Antarctic sea ice coverage.

Microwave radiometry, however, is implemented less frequently; this technique is used mostly for the global monitoring of atmospheric parameters and cloudiness, sea surface temperature and (recently) salinity, the near-surface wind vector, and sea ice, and sometimes for the local control of oil pollutions in the sea.

It has been assumed in the past years that possibilities of passive microwave radiometry and imagery for studying *submesascale* (~1–10 km) and even *mesascale* (~10–100 km) ocean dynamic processes and events are limited. There are two main reasons for that: the first is poor instrument resolution and low signal-to-noise ratio, and the second is difficulties to convert noisy radiometric signals (raw data) into relevant geophysical picture without significant errors.

To solve the first problem and achieve high spatial resolution of microwave radiometer data, a very large antenna (at least ~10–30 m at space-based observations) is required. The second problem is solved using accurate calibration and technical validation of the radiometer system. In both cases, it is necessary to create and apply a special algorithm for the retrieval of geophysical information from raw radiometric data in accordance with the instrument specification and an observation process.

The most important step is the evaluation of the so-called signatures of the interest. An efficient thematic retrieval algorithm should operate with advanced data/image processing and computer vision techniques and also invoke theoretical or (semi-)empirical models or multiparameter approximations. The creation of such a combined algorithm is a complicated scientific task, especially in the case of high-resolution microwave measurements.

Today, only one remote sensing method remains for the quantitative direct observation of ocean surface features: fine-resolution (few meters and better) airspace optical imagery. Unlike radar or radiometer data, the high visual image quality of satellite optical systems provides detailed information about surface wave processes and their spatial characteristics. However, such optical data are not always readily available for the public.

Nevertheless, a combination of high-resolution active/passive microwave and optical techniques (including infrared bands as well) is the most efficient observation strategy at the moment. Multisensor systems can provide systematic operational control of the world's oceans, embracing scale measurements from a dozen centimeters to several kilometers.

1.3 Data Processing, Analysis, and Interpretation

Data processing, analysis, and interpretation are important subdivisions in remote sensing. There is a large number of literature resources on this subject. Because we are focusing on high-resolution microwave observations, the application of enhanced image/data processing is required in order to select and extract the relevant information.

In the case of radiometry and imagery, the processing provides a representation of raw (imaging) data in usable and information (mapping) formats that is necessary for conducting geophysical research. A follow-up analysis reveals and specifies the properties and content of the collected data sets; interpretation provides physics-based insights into data mining with the goal to investigate possible effects and/or signatures. For example, geophysical representation and specification of the ocean microwave imaging data can be performed digitally using computer vision algorithms. This gives us realizations in the form of the so-called radiometric portrait.

Data processing is divided into two parts: preprocessing and actual or thematic processing. Preprocessing is used for initial formatting, correction, noise reduction, restoration, normalization, sorting, storing, and visualization of remotely sensing data.

In ocean remote sensing, thematic processing is applied for the selection, specification, and evaluation of the signatures of interest related to certain phenomena/events. Indeed, thematic processing of ocean microwave data is a part of the geophysical information system (GIS).

Ocean GIS consists of three parts: (1) input data collection, (2) data storing and management, and (3) output product (visual realizations, maps, and materials). Ocean GIS can be organized using available data/image processing algorithms which should be adapted to the specific (low-contrast and noisy) dynamic radiometric signals and measurements.

In our case, it is convenient to consider two categories of digital image processing: statistical (global) and structural (local). The first includes spectral, correlation, cluster, fractal, texture, and fusion methods. The second is intended for more detailed specification and characterization of the selected image regions, features, or elements. An enhancement, segmentation, binarization, texturization, morphological (feature's shape, size, orientation) measurements, spatial and color filtering, and some algorithms of computer vision can be applied as well.

Interpretation is based on extended knowledge of several disciplines: the observation technology, applied physics, methods of numerical modeling, simulation, and classification of microwave data. Quantitative interpretation of ocean microwave data is a complicated repetitive process that is not uniquely determined; it involves many computer science and software products. At the same time, we believe that such a combined theoretical–experimental (*data assimilation*) approach is the most comprehensive option to achieve our goals and objectives.

1.4 Theoretical Aspect

A large number of theoretical (analytical) studies and model calculations of the sea surface microwave propagation characteristics, scattering and emission, has been performed by many authors during the past several decades. In most works, microwave radiation from the ocean and atmosphere is estimated at selected electromagnetic frequencies and fixed view angles that is motivated by observation missions and instrument specifications.

Meanwhile, theoretical hydrodynamic and electromagnetic models of the ocean–atmosphere interface play a key role in data interpretation and application. One part of the electromagnetic wave theory considers scattering and emission from a rough random ocean-like surface with different statistical properties. The most well-known is the so-called two-scale model describing the contributions from both small-scale and large-scale surface irregularities independently. The other part of the theory explains the effects of microwave emission from nonuniform ocean disperse media such as foam, whitecap, bubbles, spray, and dense aerosol. This theory operates with dielectric mixing models, wave propagation models, and/or radiative transfer equation.

We investigate both parts separately and integrate them into a *composite statistical multifactor microwave model*. A composite model allows us to provide flexible analysis of spectral and polarization characteristics of ocean emissivity at variable conditions and in a wide range of electromagnetic wavelengths from 0.3 to 30 cm.

Modeling and simulations of complex microwave radiometric data (signals, images, signatures) is also an important part of advanced research. This new approach provides prediction and investigation of ocean microwave signatures through computer experiments. A number of realistic scenes and scenarios related to different oceanic processes, phenomena, or events can be investigated numerically.

Electromagnetic models of microwave emission and scattering from the sea surface are constantly updated and improved. An ultimate method or tool suitable for theoretical analysis of ocean microwave data has not yet been found (unlike, for example, Earth's land microwave observations). The parameters of the existing models and approximations are usually adjusted in order to complete the best fit for the given experimental data set. Actually, such a "modeling volatility" or kind of nonrobust connection between theory and experiment just demonstrate quite objectively the overall difficulties and challenges of ocean exploration using passive microwave methods.

1.5 Historical Chronology

The first microwave radiometer-receiver was introduced by the American physicist Robert H. Dicke in 1946 in the Radiation Laboratory of Massachusetts Institute of Technology. This radiometer operated at a wavelength of 1.25 cm and was intended to measure the temperature of environmental microwave radiation. Later, in the 1950s and 1960s, numerous microwave radiometers were designed and employed in radio astronomy, atmospheric and terrestrial studies, and also in planetary mission (Mariner 2 Venus Flyby, 1962).

The first launch of a passive microwave radiometer for Earth observation was accomplished by the U.S.S.R. Cosmos 243 satellite in 1968. This four-channel microwave radiometer at wavelengths of 0.8, 1.35, 3.4, and 8.5 cm with horn antenna and spatial resolution about 20 km provided global observations of the ocean, atmosphere, and sea ice. Then, in 1970, the Cosmos 384 satellite with the same set of microwave radiometers was launched with the same purposes.

Below is a short list of the past and current spacecraft missions operated with passive microwave radiometers and imagers and dedicated to the monitoring of ocean and atmosphere:

NASA Nimbus-5/ESMR (1972); Skylab/S 193 (1973); Nimbus-6/ESMR (1975); DMSP/SSM/T (1978); NOAA TIROS-N/MSU (1978); SEASAT 1/SMMR (1978); Nimbus 7/SSM/R (1978); Cosmos 1076 and 1151 (1979); Salyut-6/KRT-10 (1979); NOAA-7/AVHRR/2 (1981); Kosmos-1500 (1983); DMSP/SSM/I (1987); ADEOS/NSCAT (1996); NASA/JAXA TRIMM TMI (1997); Mir-Priroda/IKAR (1997); DMSP/AMSU-A/B (1998); METEOR-1/ MIMR (1998); NOAA-15/AVHRR/3 (1998); ADEOS II/AMSR (1999); EOS-PM/MIMR (2000); Meteor-3M-1/MTVZA (2001); Aqua/AMSR-E (2001); Coriolis/WindSat (2003); ESA SMOS (2009); Meteor-M No.1/ MTVZA-GY (2009); NASA Aquarius (2011); Meteor-M No.2/MTVZA-GY (2014); NASA/SMAP (2015).

Additionally, since the 1970s, passive microwave radiometers were used at different aircraft laboratory platforms: NASA Convair 990; Soviet Ilyushin Il-18 and Il-14; Antonov An-2, An-12, and An-30; Tupolev Tu-134 SKh; NASA P-3 and NRL P-3 Orion; NASA DC-8; C-130 Hercules; NOAA WP-3D; Convair-580; Dornier 228; Short Skyvan.

Radiometric measurements were conducted around the globe from ship platforms in the 1980s. In 1992, a multifrequency set of microwave radiometers and scatterometers has been installed and operated at the gyrostabilized platform of the research vessel *Akademik Ioffe* during the Joint U.S./Russia Remote Sensing Experiment JUSREX 1992 (Atlantic Ocean).

Detailed field radiometric experiments were conducted from the stationary sea platforms (WISE 2000 and 2001 Mediterranean Sea; CAPMOS 2005 Black Sea) and from the blimp (COPE 1995). Many test experiments and precise microwave radiometric measurements were performed in open air laboratory water tanks and also in natural research pool (Krylov State Research Centre, Saint Petersburg, in the 1980s).

All these programs, data, and results bring unique and remarkable experiences providing great insight into the problem. The material collected by many authors over the years allows us to realize much better the potential and benefits of passive microwave observation technology. In particular, our remote sensing experiments (1997–2004) and collected data have shown excellent capabilities of high-resolution multiband microwave imagery for observations of ocean surface features. This mission has been the most significant innovation in our studies.

1.6 Objectives of This Book

This book addresses the fundamentals of passive microwave remote sensing of ocean environment. The attention is focusing on detailed description of the physical principles, methodology, theory, and practice of ocean microwave observations. This book also offers a first look at microwave detection capabilities. In order to realize the problem as a whole, we emphasize a number of important scientific topics and results related to hydrodynamics, electrodynamics, physics-based modeling, data analyses, and interpretation.

This book has the following objectives:

- Develop and update general science, technology, and information knowledge to provide advanced remote sensing studies of ocean environment
- Provide insight into the research and specification of hydrodynamic and electromagnetic effects, contributions, and signatures potentially observable by passive microwave radiometric sensors
- Investigate and demonstrate the capabilities and advantages of high-resolution multiband passive radiometry and imagery for the detection of ocean variables and dynamic features

The material presented in this book is the radiometric part of a *multisensor synergy observation concept*. This concept includes a combined simultaneous use of active/passive microwave and optical techniques for advanced remote sensing of the ocean. It is created on the basis of our multiyear experiences, an experimental effort, and analysis of existing data, materials, publications, reports, and documents available from various literature sources.

As a result of the past and recent studies in which the author was involved continuously (1975–2015), a scientifically applied topic named *Radio-Hydro-Physics* (this terminology was used in the early 1980s by Professor V. Etkin, 1931–1995) got a fresh start in the late 1990s. One part of the scientific research has been separated under the name *Electrodynamics of the Ocean–Atmosphere Interface* (1998).

References

Mather, P. M. and Koch, P. 2011. *Computer Processing of Remotely-Sensed Images: An Introduction*, 4th edition. Wiley-Blackwell, UK.

Ulaby, F. T. and Long, D. G. 2013. *Microwave Radar and Radiometric Remote Sensing*. University of Michigan Press, Ann Arbor, Michigan.

Bibliography

Bass, F. G. and Fuks, I. M. 1979. *Wave Scattering from Statistically Rough Surfaces*. Pergamon, Oxford, UK.

Cherny, I. V. and Raizer, V. Y. 1998. *Passive Microwave Remote Sensing of Oceans*. Wiley, Chichester, UK.

Fung, A. K. and Chen, K.-S. 2010. *Microwave Scattering and Emission Models for Users*. Artech House, Norwood, MA.

Grankov, A. G. and Milshin, A. A. 2015. *Microwave Radiation of the Ocean-Atmosphere: Boundary Heat and Dynamic Interaction*, 2nd edition. Springer, Cham, Switzerland.

Ishimaru, A. 1991. *Electromagnetic Wave Propagation, Radiation, and Scattering*. Englewood Cliffs, Prentice Hall, New Jersey.

Kramer, H. J. 2002. *Observation of the Earth and Its Environment: Survey of Missions and Sensors*, 4th edition. Springer, Berlin, Germany.

Kraus, J. D. 1986. *Radio Astronomy*, 2nd edition. Cygnus-Quasar Books, Powell, Ohio.

Lavender, S. and Lavender, A. 2015. *Practical Handbook of Remote Sensing*. CRC Press, Boca Raton, FL.

Martin, S. 2014. *An Introduction to Ocean Remote Sensing*, 2nd edition. Cambridge University Press, Cambridge, UK.

Matzler, C. 2006. *Thermal Microwave Radiation: Applications for Remote Sensing*. The Institution of Engineering and Technology, London, UK.

Njoku, E. G. 2014. *Encyclopedia of Remote Sensing (Encyclopedia of Earth Sciences Series)*. Springer, New York, NY.

Pratt, W. K. 2007. *Digital Image Processing*, 4th edition. John Wiley & Sons, Hoboken, New Jersey.

Robinson, I. S. 2010. *Discovering the Ocean from Space: The Unique Applications of Satellite Oceanography*. Springer, Berlin, Germany.

Rytov, S. M., Kravtsov, Yu. A., and Tatarskii, V. I. 1989. *Principles of Statistical Radiophysics*. Vol. 3. Springer-Verlag, Berlin.

Sharkov, E. A. 2003. *Passive Microwave Remote Sensing of the Earth: Physical Foundations*. Springer Praxis Books, Chichester, UK.

Skou, N. and Le Vine, D. M. (2006). *Microwave Radiometer Systems: Design and Analysis*, 2nd edition. Artech House, Norwood, MA.

Ulaby, F. T., Moore, R. K., and Fung, A. K. 1981, 1982, 1986. *Microwave Remote Sensing. Active and Passive* (in three volumes), Advanced Book Program, Reading and Artech House, Norwood, MA.

Voronovich, A. 1999. *Wave Scattering from Rough Surfaces (Springer Series on Wave Phenomena)*, 2nd edition. Springer, Berlin, Heidelberg.

Woodhouse, I. H. 2005. *Introduction to Microwave Remote Sensing*. CRC Press, Boca Raton, FL.

2

Ocean Phenomena

2.1 Introduction

Today, there is considerable interest in achieving better performance of ocean observations using active/passive microwave remote sensing techniques. In order to extract the geophysical information from the collected microwave data, it is necessary not only to understand the mechanisms of electromagnetic wave propagation—scattering and emission from the ocean surface, but also to learn geophysical processes and phenomena occurring at the air–sea interface. Several great books (Lamb 1932; Kitaigorodskii 1973; Phillips 1980; Craik 1985; Apel 1987; Kraus and Businger 1994; Miropol'sky 2001; Janssen 2009) provide comprehensive and needed information on physical oceanography and hydrodynamics.

The goal of this chapter is to give the reader an initial knowledge base about the main oceanic phenomena and hydrodynamic factors, which are responsible for the formation and variations of microwave remotely sensed data. The emphasis is to specify processes and events that are potentially observable by passive (and active) microwave sensors. The selected material presented below is also intended for researchers and specialists who are interested in developing and/or conducting complex hydrodynamic studies involving microwave and other remote sensing measurements.

2.2 Structure of the Ocean–Atmosphere Interface

Figure 2.1, initially created in the Woods Hole Oceanographic Institution, California and updated here in more detail, demonstrates a number of phenomena in the ocean that are important for microwave diagnostics. Upon considering this picture, it becomes clear that the complete description of the real-world ocean environment is an extremely challenging task, which, perhaps, may not be solved using conventional theories and/or analytic approaches.

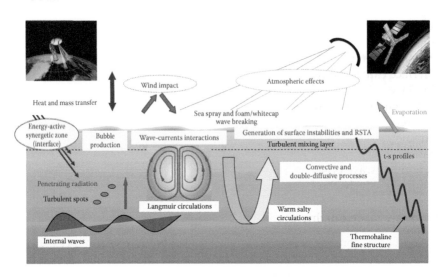

FIGURE 2.1

Ocean environment in a view of microwave remote sensing. (Based on illustration by Jayne Doucette, Woods Hole Oceanographic Institution, California.) Pictures: Left upper corner—Satellite NASA Aquarius. Right upper corner—Satellite ESA SMOS.

Briefly, the structure of the ocean–atmosphere interface can be divided into three categories: the near-surface upper ocean layer, the interface itself, and atmospheric boundary layer. The upper ocean layer is characterized by thermohaline finestructure, double-diffusive convection, circulations, internal wave motions, and turbulence. The near-surface atmospheric boundary layer (with thickness ~10 m above the surface) is characterized by turbulent fluxes, stratification, and stability.

Because most remote sensing observations provide statistical and averaged data, satellite scientists-oceanographers usually operate with semiempirical models and approximations in order to investigate large-scale dynamics of the ocean surface, wind-generated waves, fluxes, and boundary-layer parameters.

In addition to such a global geophysical interpretation, the author proceeds from the assumption that the microwave response from the ocean surface is defined by many individual structural and dynamic factors and localized processes. Therefore, from our point of view, the ocean–atmosphere interface should be described as a stochastic multiscale dynamic system with a large number of distributed hydro-physical parameters and multiple interconnections.

The borderline between ocean and atmosphere, which is shown graphically in Figure 2.1, has much more complicated and ambiguous internal content. This interface includes microlayers of organic and nonorganic surfactants, turbulent mixing macrolayers, as well as multiscale geometrical and volume nonuniformities, which are, eventually, surface waves and foam/whitecap coverage. Both these geometrical and volume nonuniformities produce significant impacts on microwave radiometric measurements.

More understandable "remote sensing" definition of the ocean–atmosphere interface can be formulated in terms of electrodynamically stratified transition layer of variable structure, configuration, and thickness. An adequate and accurate microwave analysis of such a composition interface layer will require more detailed investigation and knowledge of distinguished parameters of individual components of overall system rather than its statistical-based averaged characteristics. Moreover, in order to build an efficient microwave observation (detection) technology, we also have to explore comprehensively the "behavior" of the ocean surface (background) at different conditions. Below, we consider the main hydrodynamic factors and processes related to this problem.

2.3 Classification of Surface Waves

Ocean surface waves are strongly diversified by geometrical form and space–time scales. For this reason, two approaches are used to describe them. The first, deterministic, is based on the application of the fundamental hydrodynamic theory. It describes the configuration profile (shape) of a regular linear or nonlinear wave on deep or shallow water. The second, statistical, operates with the probabilistic laws of distribution of energy between different wave components. In this case, it is presumed that the surface elevation fluctuates randomly in space and time and can be described as a statistical ensemble of a large number of surface harmonics.

More adequate methods that unite both approaches are connected with numerical solutions of hydrodynamic and energy balance equations. The most important result of numerical methods is the definition of numerical profiles of two-dimensional and even three-dimensional nonlinear surface waves and the modeling of their evolution in space and time up to the moment of breaking. It is also possible to investigate the phenomenon of the bifurcation of gravity waves due to their interactions and establish the criteria of instability.

In hydrodynamics, steady- and nonsteady-state surface waves are distinguished. Steady-state waves do not change their properties in space and time. Otherwise the waves are named nonsteady-state waves. In addition, periodical linear and nonlinear steady surface waves are separated (Table 2.1).

An important type is the surface gravity waves of finite amplitude (Stokes waves). These waves are unsteady with respect to small periodic disturbances (Benjamin–Feir modulation instability). The effects of instability and evolution and bifurcations of one-dimensional and two-dimensional

TABLE 2.1

Classification of Steady-State Surface Waves (by Theory)

Type of Surface Waves	Author
1. Linear periodical	Nekrassov (1951)
2. Trochoidal	Gerstner (1802)
3. Nonlinear periodical of finite amplitude	Stokes (1847)
4. Gravity solitary (soliton)	Boussinesq (1890)
5. Capillary linear periodical	Sekerzh-Zenkovich (1972)
6. Capillary nonlinear periodic	Crapper (1957)
7. Capillary solitary (soliton)	Monin (1986)

Source: Cherny I. V. and Raizer V. Yu. *Passive Microwave Remote Sensing of Oceans.* 195 p. 1998. Copyright Wiley-VCH Verlag GmbH & Co. KGaA. Reproduced with permission.

surface nonlinear gravity waves in deep water have been investigated in detail (Zakharov 1968; Yuen and Lake 1982; Craik 1985; Su 1987; Su and Green 1984).

Flat weakly nonlinear waves are described by the Korteweg–de Vries equation (1895). The solutions of this equation can be as periodical as solitary waves (solitons). The existence of gravity–capillary solitons in shallow water was proved theoretically on the basis of the nonlinear Kadomtsev–Petviashvilli and Schrödinger equations.

Capillary waves or ripples are essentially nonlinear. The theoretical profile of capillary waves has a complex and ambiguous geometrical form. Short capillary waves in the ocean are strongly unsteady. Although they are not regular waves in the classical hydrodynamic sense, they can be represented by a random field of surface impulse-type perturbations of high steepness.

Finally, we include a category named "turbulent roughness" or microscale surface turbulence. This category represents a nonsteady field of small-scale, randomly distributed on the surface fluctuating disturbances. Such disturbances occur under the influence of boundary-layer turbulent flows, microbreaking processes, local variations of the near-surface winds, strong (sub) surface currents, or as result of interaction of water droplets (from spray or rain) with the ocean surface. The contribution of turbulent roughness to ocean microwave emission cannot be neglected at observations concerning the nature of surface-active films, local variations of sea surface temperature and salinity, or turbulent wakes.

2.4 Generation and Statistics of Wind Waves

An ensemble of wind-generated surface waves is the main environmental factor in remote sensing, which should be carefully investigated. Wind waves represent multiscale dynamic geometric disturbances, generated mostly stochastically through multiple cascade processes. Sometimes, we may separate environmental wind waves and other possible (induced) surface disturbances because there are some differences in their generating mechanisms and geometrical properties. But it is quite difficult to distinguish them properly by spectral and statistical characteristics. Therefore, adequate hydrodynamic description and modeling of an overall system of ocean surface waves, their scales, evolution, and dynamics are still of great interest in many applications.

2.4.1 Generation Mechanisms

The well-known surface wave generation mechanisms are the following:

- Surface wind stress
- Kelvin–Helmholtz instability due to local wind shear
- Miles shear instability due to the influence of a matching layer with wind profile
- Resonance mechanism due to nonlinear interactions of gravity waves when the speeds of wave propagation and wind are the same (Phillips 1980)
- Weak turbulence theory (Zakharov and Zaslavskii 1982) due to the locality of wave–wave interaction in the case of a wind-driven sea

In past years, the approach of slow dispersion and nonlinearity of deterministic surface gravity–capillary waves has been developed. Using this theory, new solutions of the Korteweg–de Vries equation concerning dynamics of solitons and their interactions were investigated (Craik 1985).

Another mechanism of surface wave generation deals with nonlinear wave–wave interaction. The dynamics of the interaction are described by the kinematic theory for statistical ensemble of surface waves (Hasselman 1962). This theory also describes the formation of wave number spectrum in the oceans. An important application of the theory is the consideration of surface *wave–current* interactions. In particular, the effects of blocking gravity–capillary waves by surface currents induced by internal waves have been manifested (Section 2.4.7). As a result, the strong transformation of the surface wave number spectrum in the interval of decimeter surface wavelengths occurs (Basovich and Talanov 1977).

In the case of linear theory, the amplitude of surface waves decrease when they are propagated along the current, but the amplitude increases when waves are propagated against the current. It is possible to register both an increase and decrease of the wave energy's spectral density using radar observations. An example is the propagation of surface waves on horizontally nonuniform current in the field of oceanic internal waves. The theory explains the effects of the occurrence of anomalous roughness such as surface smoothing, slicks, and "rip currents."

2.4.2 Statistical Description and Wave Number Spectrum

The statistical description is based on integral information about the change of averaging spectral density of wave energy, only due to slow variations of wind speed and interaction between air flow and ocean surface. Moreover, the wave field in the ocean is a multiple-scale nonlinear dynamic system, which is characterized by a large degree of freedom. Resonance and nonresonance groups of waves exist in such a system.

As it follows from a general theory, wave–wave interactions provide a stable spatial evolution of the system. But resonance wave–wave interactions under certain conditions lead to the generation of different hydrodynamic instabilities, which with time lead to chaotic surface motions. In order to predict and model the behavior of such a dynamic wave system in space and time, spectral-based mathematical formalism is used.

The exact universal all-purpose formula, which describes two-dimensional wave number spectra of ocean surface waves in the wide range of spatial frequencies, does not exist. There are empirical and theoretical approximations of surface wave spectra (Phillips 1980; Pierson and Moskowitz 1964; Mitsuyasu and Honda 1974; Leikin and Rosenberg 1980; Mitsuyasu and Honda 1982; Keller et al. 1985; Merzi and Graft 1985; Phillips and Hasselmann 1986; Donelan and Pierson 1987; Komen et al. 1996; Engelbrecht 1997; Young 1999; Mitsuyasu 2002; Lavrenov 2003; Janssen 2009; Kinsman 2012) that are used in remote sensing. According to these and other data, the energetic part of full wave number spectrum can be separated on the following five regions:

1. Region of large energy-carrier quasi-linear gravity waves (the Pierson–Moskowitz spectrum):

$$F_1(K) = 4.05 \cdot 10^{-3} K^{-3} \exp\left\{\frac{-0.74g^2}{[V^4(u_*)K^2]}\right\} \tag{2.1}$$

for interval $0 < K < K_1 = K_2 u_{*m}^2 / u_*^2$
V is the wind speed at an altitude of 19.5 m (m/s)
$u_* = \sqrt{C_n V^2}$ is the friction velocity (cm/s)
$C_n = (9.4 \cdot 10^{-4} V + 1.09) \cdot 10^{-3}$ is the aerodynamic coefficient of drag
$u_{*m} = 12$ cm/s

2. Region of nonlinear short gravity waves:

$$F_2(K) = 4.05 \cdot 10^{-3} K_1^{-1/2} K^{-5/2} \tag{2.2}$$

for equilibrium interval $K_1 < K < K_2 \approx 0.359$ cm^{-1}.

3. Transfer region of dynamical equilibrium:

$$F_3(K) = 4.05 \cdot 10^{-3} D(u_*) K_3^{-\rho} K^{-3+\rho}, \tag{2.3}$$

$\rho = \log[u_{*m} D(u_*)/u_*]/\log(K_3/K_2),$
$K_2 < K < K_3 \approx 0.942 \ \text{cm}^{-1},$

where

$$D(u_*) = (1.247 + 0.0268 u_* + 6.03 \cdot 10^{-5} u_*^2)^2 \tag{2.4}$$

(the Pierson and Stacy approximation); or

$$D(u_*) = 1.0 \cdot 10^{-3} u_*^{9/4} \tag{2.5}$$

(the Mitsuyasu and Honda approximation).
   Another form is

$$F_3(K) = F_4(K_3) \left(\frac{K}{K_3} \right)^q, \tag{2.6}$$

$K_2 < K < K_3 \approx 0.942 \ \text{cm}^{-1},$

$$q = \frac{\log[F_2(K_2)/F_4(K_3)]}{\log(K_2/K_3)},$$

where $F_4 (K_3)$ corresponds to Equation 2.7.
4. The equilibrium range of the Phillips' spectrum of limiting gravity–capillary waves:

$$F_4(K) = 4.05 \cdot 10^{-3} D(u_*) K^{-3}, \quad K_3 < K < K_v, \tag{2.7}$$

$$K_v = 0.5756 u_*^{1/2} [D(u_*)]^{-1/6} K_m,$$

$$K_m = \left(\frac{\rho_w g}{\gamma_0} \right)^{1/2} \cong 3.63 \ \text{cm}^{-1},$$

where g is gravity, ρ_w is density of the water, and γ_0 is the surface tension coefficient.

Another form is

$$F_4(K) = 0.875(2\pi)^{\rho_1-1}\frac{g+3gK^2/13.1769}{(gK+gK^313.1769)^{(\rho_1+1)/(2)}},\qquad(2.8)$$

$K_3 < K < K_4,$
$\rho_1 = 5.0 - \log u_*,$

where K_4 is defined from equation

$$F_4(K_4) = F_5(K_4).\qquad(2.9)$$

5. Region of capillary waves and weak turbulence:

$$F_5(K) = 1.479\cdot10^{-4}u_*^3K_m^6K^{-9},\qquad(2.10)$$

$$K_v < K < \infty$$

Full wave number spectrum F(K,V) calculated using Equations 2.1 through 2.10 is shown in Figure 2.2. The spectrum is parameterized by wind speed (V).

There are a few more spectral models and approximations related to wind-generated surface waves in wide frequency intervals (Huang et al. 1981;

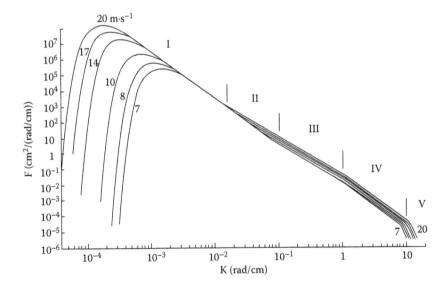

FIGURE 2.2
Wave number spectra at variable conditions. Five spectral intervals are combined all together according to Equations 2.1 through 2.10. Wind speed value is marked from 7 to 20 m/s. (Cherny I. V. and Raizer V. Yu. *Passive Microwave Remote Sensing of Oceans*. 195 p. 1998. Copyright Wiley-VCH Verlag GmbH & Co. KGaA. Reproduced with permission.)

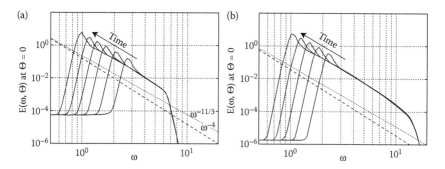

FIGURE 2.3

Self-similar wave spectra based on the Zakharov's theory. Nondimensional energy spectral density versus nondimensional frequency. Temporal evolution of the spectrum during several hours is shown by the arrow. Exponential asymptotes: (dash) "–4" and (dotted) "–11/3". Wind speed: (a) 10 m/s and (b) 20 m/s. (Adapted from Badulin, S. I. et al. 2005. *Nonlinear Processes in Geophysics*, 12:891–945.)

Glasman 1991a,b; Apel 1994; Romeiser et al. 1997; Kudryavtsev et al. 1999; Hwang et al. 2000a,b; Plant 2015). Many researchers refer to the spectral model (Elfouhaily et al. 1997), which is supposed to be a best-fit spectrum for polarimetric radar observations. The Elfouhaily spectrum is based on hydrodynamic properties of the sea surface and describes wind dependencies reproduced from the Cox–Munk slope distributions (Cox and Munk 1954).

We also refer to Figure 2.3 which illustrates computed self-similar wave spectra based on Zakharov's theory (Badulin et al. 2005). This spectrum is defined numerically from the solution of the energy balance equation. It seems to be the most suitable physics-based theoretical model of dynamic wave spectrum needed for advanced remote sensing studies.

2.4.3 Surface Dynamics: Elements of Theory

The fundamental description of atmosphere and ocean dynamics is based on the Navier–Stokes equations:

$$\frac{\partial \vec{V}}{\partial t} + (\vec{V} \cdot \nabla)\vec{V} = -\frac{1}{\rho_t}\nabla\vec{P} + v_0\nabla^2\vec{V} + \vec{F}, \tag{2.11}$$

$$\nabla \cdot \vec{V} = 0,$$

where $\vec{V}$ is the velocity vector and $\vec{P}$ is the pressure vector at each point $\vec{r}$ and instant t; ρ_t is the fluid density; v_0 is the kinematic viscosity; and $\vec{F}$ is the forces term (gravity, stirring). Usually, the solid boundaries or free surface and fluid boundaries are considered. Therefore, in the common case, both nonlinear kinematic and dynamical boundary conditions may introduce together with Equation 2.11.

Investigations of the nonlinear Equation 2.11 show that two principal types of solutions can be found—stable and unstable. This means that in the classic understanding, "real motions must not only satisfy the equations of hydrodynamics, but must be stable in the sense that the perturbations which inevitably arise under actual conditions must die out with time" (Monin and Yaglom 2007). It is clear that such a suggestion imposes stringent limits on a relation of an initial medium parameter and nonlinearity degree of the equations. Usually, the Reynolds number $Re = U_0L/\nu_0$ (where U_0 is the characteristic velocity, L is the characteristic scale, and ν_0 is the kinematic viscosity) is applied as the main criterion of the stability. The stable or unstable regimes of motion are determined by the value of the critical Reynolds number Re_c. If $Re < Re_c$, the regime is stable; if $Re > Re_c$, the regime can be unstable. Also, this criterion is used to estimate the ratio of the nonlinear terms to the dissipative terms in the Navier–Stokes equation.

In the common case of wave–wave and wave–current interaction, the evolution of the spectral density of wave energy is described by the kinetic equation (Hasselman 1962):

$$\frac{\partial N}{\partial t} + (\vec{U} + \vec{C}_g)\nabla N = I_{in} + I_{n1} + I_{ds}, \tag{2.12}$$

where $N(\vec{K}, \vec{r}, t) = \rho\omega_0/|\vec{K}|S(\vec{K}, \vec{r}, t)$ is the action spectral density, $\vec{C}_g$ is the local group velocity, $\vec{U}$ is the current velocity vector, and $S(\vec{K}, \vec{r}, t)$ is the two-dimensional wave number spectrum.

The processes, which modify the action spectral density, are described by the net source function $I_s = I_{in} + I_{nl} + I_{ds}$ on the right side of the equation. The source function is represented as the sum of the three terms: the energy flux from the wind to wave I_{in}; the energy flux due to nonlinear resonance wave–wave interactions I_{nl}; and the energy loss due to wave breaking and other dissipative processes I_{ds}. In the case of the nonuniform surface current field, induced, for example, by internal wave packets, the dispersion relation for the surface waves may be written as

$$\omega(\vec{K}, \vec{r}, t) = \omega_0(\vec{K}, \vec{r}) + \vec{K}\vec{U}, \tag{2.13}$$

where $\omega_0(\vec{K}, \vec{r})$ is the dispersion relation for initial (nondisturbed) surface waves.

The spectral function of perturbation associated with any hydrodynamic process (for example, internal waves or surface currents) may be written as

$$f(\vec{K}, \vec{r}, t) = \frac{S_f(\vec{K}, \vec{r}, t) - S(\vec{K})}{S(\vec{K})}, \tag{2.14}$$

where $S(\vec{K})$ is the initial (nondisturbed) wave number spectrum.

On a base of Equations 2.11 through 2.14, in principle, it is possible to calculate the perturbation spectrum $S_f(\vec{K}, \vec{r}, t)$ and use it as an input parameter in the ocean microwave models and applications. This way was used for quantitative analysis of radar signatures of surface waves and internal waves in the field experiments JOWIP and SARSEX (Gasparovic et al. 1988; Thompson et al. 1988). Obviously, similar description may also apply for the interpretation of radiometric microwave signatures, associated with the influence of surface roughness disturbances.

Analysis of the action balance equation was carried out by many authors (Phillips 1980; Zakharov and Zaslavskii 1982; Zaslavskiy 1996). In particular, problems of surface modulation and surface wave–current interaction in the field of internal waves were investigated. However, only simple approaches, when the source function in (2.12) equaled $I_s = 0$, or, $I_s = I_{in}$, or $I_s = I_{nl}$, were considered in detail. Moreover, the universal character of full dynamic nonlinear Equation 2.11, and Equation 2.12, permit modeling different oceanic scenes and scenarios numerically, including the generation of dynamic surface structures and instabilities.

2.4.4 Surface Wave–Wave Interactions and Manifestations

Nonlinear wave–wave interactions can be separated into two types: weak and strong. The first type of synchronous interactions are first-order nonlinear effects for surface waves of finite amplitude with relatively small slope. Nonlinearity causes a slow change of wave characteristics in the space and time and provides small perturbations. This process is characterized by a long duration of interactions. The second type is characterized by small time and small spatial scales of interactions. In this case, different types of instabilities are advanced. The strong interactions cause, for example, the wave breaking phenomena.

For second-order resonance interactions among a triad of surface waves, the following conditions of synchronism must be satisfied simultaneously:

$$\vec{K}_1 = \vec{K}_2 + \vec{K}_3, \quad \omega_1 = \omega_2 + \omega_3, \quad \omega = (gK)^{1/2}, \tag{2.15}$$

where K and ω are the wave number and wave frequency. There are no nontrivial solutions of Equation 2.15. But resonance cannot occur at this order, and only the effect of the perturbation of the wave profile can be seen (Phillips 1980).

The interaction of the three wave components $(\vec{K}_1, \vec{K}_2, \vec{K}_3)$ at the quadratic and cubic orders generate the components with the numbers $(\vec{K}_1 \pm \vec{K}_2 \pm \vec{K}_3)$. For resonance among a tetrad of wave components, the conditions of synchronism must be or near satisfied,

$$\vec{K}_1 \pm \vec{K}_2 \pm \vec{K}_3 \pm \vec{K}_4 = 0, \quad \omega_1 \pm \omega_2 \pm \omega_3 \pm \omega_4 = 0, \quad \omega = (gK)^{1/2}. \tag{2.16}$$

The nontrivial solution of Equation 2.16 exists for four-wave interactions:

$$\vec{K}_1 + \vec{K}_2 = \vec{K}_3 + \vec{K}_4, \quad \omega_1 + \omega_2 = \omega_3 + \omega_4, \quad \omega = (gK)^{1/2}. \tag{2.17}$$

This scheme describes the four-wave interactions of weakly nonlinear surface gravity waves in deep water (Zakharov 1968). This interaction mechanism causes energy transfer in the space–time spectrum, and effects its broadening at wind-waves-generating conditions (Hasselman 1962).

There is an important and particular case of a four-wave interaction model when two of the primary wave numbers are coincident ($\vec{K}_3 = \vec{K}_4$). The resonance conditions (2.17) change as

$$\vec{K}_1 + \vec{K}_2 = 2\vec{K}_3, \quad \omega_1 + \omega_2 = 2\omega_3. \tag{2.18}$$

These conditions were tested and investigated experimentally in a laboratory when wave number vectors $\vec{K}_1$ and $\vec{K}_2$ were perpendicular (Phillips 1980). But under open ocean conditions, strict satisfaction of the resonance for several systems of surface waves is impossible. The phenomenon of quasi-synchronism due to nonstationary and noncoherent interaction between weakly nonlinear gravity waves was investigated, using satellite, airborne radar, and optical remote sensing data (Beal et al. 1983; Grushin et al. 1986; Volyak et al. 1987; Raizer et al. 1990; Raizer 1994; Voliak 2002).

In the ocean, it is possible to observe quasi-resonance wave components which satisfy the conditions:

$$\vec{K}_1 + \vec{K}_2 = \vec{K}_3 + \vec{K}_4 - \Delta\vec{K},$$

or

$$\vec{K}_1 + \vec{K}_2 = 2\vec{K}_3 - \Delta\vec{K}, \quad \text{or} \quad 2\vec{K}_1 - \vec{K}_2 = \vec{K}_3 + \Delta\vec{K}, \tag{2.19}$$

where $\Delta\vec{K}$ is the phase mismatch. The value of the phase mismatch characterizes the group structure of interacting waves and depends on the extent of nonstationary or nonuniformity of the investigated wave-generating system.

Earlier remote sensing investigations of large-scale wave–wave interactions in the ocean surface were conducted in 1976–1978, using the airborne side-looking radar "Toros" operated at 2.25 cm wavelength. Since 1981, methods of radar imagery and aerial photography have been applied simultaneously to study the wave group structures of wind-generated gravity waves and the dynamics of surface nonlinear wave–wave interactions (Grushin et al. 1986; Raizer et al. 1990; Voliak 2002).

Standard harmonic two-dimensional analysis of the radar and optical images in a coherent or digital processor provides information on spatial spectra and angular distributions of the wave's components. The spatial frequencies (or wave number vectors) in the regions close to spectral maximum are separated with a high degree of accuracy using digital Fourier transforms. Thus, different orientations of the vectors can be tested. For example, if synchronism between any wave components exist, then it is possible to identify the result as a nonlinear wave–wave interaction. However, the strict satisfaction of synchronism as a rule is not observed because interacted surface waves are not monochromatic.

From the radar images and aerial photography, it was clear that there were several surface wave systems oriented at different angles. The digital two-dimensional Fourier analysis was made for an exact measurement of the wave number vectors. Also, methods of low-frequency filtration and spatial averaging were applied for the accurate determination of large-scale wave components.

Figure 2.4 illustrates an example of very first manifestation of wave–wave interactions registered in the airborne radar images after their spectral analysis (Volyak et al. 1985, 1987). Three-wave systems are manifested: two basic wave systems and a third additional wave system oriented at an angle of about 30°. The diagram of the wave number vectors $\bar{K}_{10}$ (the additional

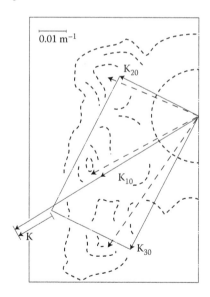

FIGURE 2.4
Experimental radar-based wave number vector diagram corresponding to four-wave interaction scheme. Dotted line arrows: data from direct analysis; solid line arrows: corrected values. The dot and dash lines complete the spatial synchronism parallelogram $K = 2K_{10} - K_{20} - K_{30}$. (Adapted from Volyak, K. I. et al. 1985. *Atmospheric and Oceanic Physics*, 21(11):895–901 (translated from Russian); Volyak, K. I., Lyakhov, G. A., and Shugan, I. V. 1987. In *Oceanic Remote Sensing*. Nova Science Publishers, pp. 107–145 (translated from Russian); Cherny I. V. and Raizer V. Yu. *Passive Microwave Remote Sensing of Oceans*. 195 p. 1998. Copyright Wiley-VCH Verlag GmbH & Co. KGaA. Reproduced with permission.)

system), $\vec{K}_{20}$, and $\vec{K}_{30}$ (two basic systems) illustrates their spatial distributions. A simple test shows that four-wave interaction scheme is satisfied:

$$\sqrt{|\vec{K}_{20}|} + \sqrt{|\vec{K}_{30}|} = 2\sqrt{|\vec{K}_{10}|}. \tag{2.20}$$

The following values of the wave numbers (or wavelengths) and angles were obtained: $|\vec{K}_{10}| = 0.033 \text{ m}^{-1}$ ($\Lambda = 190 \text{ m}$), $\alpha = 33°$; $|\vec{K}_{20}| = 0.045 \text{ m}^{-1}$ ($\Lambda = 140 \text{ m}$), $\alpha = 62°$; $|\vec{K}_{30}| = 0.025 \text{ m}^{-1}$ ($\Lambda = 250 \text{ m}$), $\alpha = -29°$; the modulus of the wave vectors mismatch is equal to $|\vec{K}| = 0.013 \text{ m}^{-1}$. In this case, "quasi-synchronous," the cubic interaction of randomly modulated surface waves was manifested.

Another example deals with experimental data obtained from SEASAT synthetic aperture radar (SAR) images (Beal et al. 1983). Although the radar images have been with a clear speckle structure that masks the wave systems, a number of original image spectra was calculated and used for testing the wave–wave interactions. It was confirmed, in particular by Volyak et al. (1987), that the "spatial synchronism geometry" of the surface wave structures satisfy four-wave interaction schemes (2.19) for different combinations of wave number vectors and modules. The distribution of the wave vectors in space depends on ocean surface conditions. This fact has a fundamental importance for detection purposes in context with active/passive and optical observations of the ocean surface.

Nonlinear wave–wave interactions can also be investigated using a high-resolution optical (in the visible range) imagery of the ocean surface. In particular, our airborne optical data (obtained during the period 1985–1992) allowed us to develop different schemes of four-wave interactions according to Equation 2.19 at nonstationary ocean surface conditions. For example, at the strong wind fetch, an evolution of spatial spectrum is accompanied by redistribution of the wave components with the wavelengths $\Lambda = 20$–40 m. A more detailed digital analysis of a large set of high-resolution optical data shows that wave–wave interactions can be measured in the region of short gravity waves ($\Lambda = 3$–5 m) as well.

Finally, combined radar/optical remote sensing observations due to different spatial resolutions and swaps yield unique information about dynamics of two- and even three-dimensional wave systems and perturbations induced by nonlinear interactions. Large-scale modulations of gravity waves can also be identified by combined radar/optical remotely sensed data with a good accuracy.

2.4.5 Weak Turbulence Theory

Understanding of ocean turbulence and its surface manifestations is critical for the development of advanced remote sensing techniques and nonacoustic methods of detection. The phenomenon is extremely complex and has been studied by many researchers over the years. Theoretical and experimental data can be found in several books and the corresponding reference sections, for example, Monin and Ozmidov (1985) and Thorpe (2005). Some elements of turbulence theory are also important for the assortment of the so-called turbulent wake and its interaction with wind waves (Benilov 1973, 1991).

Two types of ocean surface turbulence are distinguished: weak and strong turbulence. Weak turbulence is usually associated with dynamics of gravity surface waves in deep water, whereas strong turbulence occurs due to wave breaking activity. Both weak and strong turbulence may appear under the influence of nonlinear wave–wave interactions, strong currents, modulation instabilities, and/or localized hydrodynamic disturbances.

The weak turbulence theory is based on the solution of the kinetic equation for a spatial wave spectrum or the spectral density of the wave action. This equation accounts for four-wave resonance interactions if the dispersion law is of nondecay type, such as surface gravity waves, and three-wave interactions for the decay type laws like capillary waves. If a wave field is statistical isotropic, these equations have exact stationary solutions in the form of the power law known as the Kolmogorov spectrum (Monin and Yaglom 2007).

For surface gravity waves, two solutions exist in terms of the spectral density of wave energy. The first solution is

$$F(K) = \alpha q g^{-1/2} q^{1/3} K^{-7/2},\qquad(2.21)$$

where q is the energy flux down the spectrum and α is the nondimensional constant. The second solution is

$$F(K) = \alpha q p^{1/3} K^{-10/3},\qquad(2.22)$$

where p is the action flux up the spectrum. These spectra have been tested many times using oceanographic (*in situ*) measurements and some remote sensing data.

In the case of capillary waves, there are only wave–wave resonance triplets. The corresponding Kolmogorov spectrum is

$$F(K) = \frac{3}{2}\alpha \sigma q^{1/2} \sigma^{1/4} K^{-11/4},\qquad(2.23)$$

where σ is the surface tension coefficient. Note that the magnitude of the exponent in Equations 2.21 through 2.23 is smaller than value "4" (corresponding to the Phillips equilibrium spectrum).

It is important to note that the weak turbulence theory in the simplest form cannot explain the narrow angular distribution of wave energy for stationary ocean surface conditions. In this connection, the interaction of a wave field and a nonpotential mean surface current was investigated theoretically. It was found that induced spatial "scattering" of surface waves on the shear current gives a narrow angular spectrum of gravity waves. But mechanisms of formation of the angular wave spectra in the ocean are not fully studied and require additional investigations.

2.4.6 Hydrodynamic Instabilities in the Ocean

Recent progress in the theory of wave motions in fluid flows has been dramatic in the understanding of nonlinear wave dynamics. In particular, the role of multiple wave–wave interactions, secondary modulational instabilities, and amplification's mechanisms in the processes of generation of coherent hydrodynamic structures causing the fully developed turbulence and intermittency, has been studied and described (Moiseev and Sagdeev 1986; Moiseev et al. 1999; Charru 2011). One exclusive real-world environmental example of strong surface instabilities is the so-called *suloy* (which is a Russian word) or in English terminology—rip currents (Barenblatt et al. 1985, 1986a,b; Fedorov and Ginsburg 1992).

Instabilities play an important role in ocean boundary-layer dynamics causing the transition of wave motions to turbulence accomplishing by the change of surface wave spectrum. Owing to surface instabilities, the redistribution of wave energy occurs as spontaneously as through multiple cascades that can lead to the excitation or suppression of certain wave components. Theoretically, it means that the corresponding electromagnetic (microwave) response should vary as well. Although it is difficult to register narrowband short-term transformations in the wave number spectrum using passive microwave radiometers, however, the averaged deviation of microwave signal can be detected and recognized in some specific situations.

Hydrodynamic instabilities have been studied extensively (Faber 1995; Grue et al. 1996; Riahi 1996; Drazin 2002; Manneville 2010; Charru 2011; Yaglom and Frisch 2012). In fluid dynamics, the following classification of instabilities is used: (1) *primary instability* becomes possible if basic flow state changed to another flow state under the influence of its critical instability parameters; (2) *secondary instability* occurs when the flow state has been changed already due to primary instability and it is changed again due to the influence of critical instability parameters; and (3) *tertiary instability* is a result of the consecutive actions of preliminary and secondary instabilities.

At the same time, there is a number of well-known types of hydrodynamic instabilities (Faber 1995) observed in the real world, which are

- *Kelvin–Helmholtz instability* at the interface between two fluids moving with different velocities
- *Rayleigh–Taylor instability* at the interface between two fluids of different densities
- *Benjamin–Feir instability* or modulation instability for nonlinear Stokes waves on the water surface

- *Taylor–Couette instability* related to convection rolls, vortexes, and/or spiraling eddies
- *Benard instability* for many varieties of ocean–atmosphere convection
- Baroclinic instability in stratified shear flows

The analysis of secondary unstable processes forming the nonlinear pumping of energy from small to large scale is of practical interest in plasma and hydrodynamics (Moiseev et al. 1999; Rahman 2005). In this situation, there occurs self-similarity at the transition of hydrodynamic systems (for example, shear flow) from one state to other state. It is well known from the theory and experiments that the secondary instability as a possible wave generation mechanism can trigger and amplify certain harmonics in the surface wave spectrum (Craik 1985). During nonlinear interactions in the multiparameter and nonequilibrium dynamic system such as stratified turbulent flow in the ocean upper layer, self-similar growth processes and/or anomalous motions may occur and evolutionalize in space and time. Surface manifestations of self-similar hydrodynamic structures can be detected by microwave radiometers if fractal-based processing of the collected data is applied.

Extremely interesting phenomena occur at the stability threshold (or near it) for a system in which the secondary instability does not develop spontaneously. Such a state is quite typical, for example, for the regions of ocean with intense flows, several gradients of physical parameters, strong current shear, and other "critical" motions. The primary instability gets saturated as a result of nonlinear evolution. But background flows and gradients of physical parameters usually do not disappear; they only are changed if the motion falls below the stability threshold. It turns out that if certain outer action or perturbation enters into such an unstable system, its energy may increase significantly.

In recent years, a large interest has developed in the investigation of the resonance interaction of "burst"-type surface waves in stratified shear flows with discontinuous profile (Craik 1985). Such "burst"-type instability may explain the formation of turbulent spots in the ocean in the presence of stratified mean flows. The point of maximum interest lies in the processes with minimum "burst" time. The possibility of the existence of such rapidly growing "burst" solutions for turbulent shear flows is not obvious *a priori* but the predictions have been made from numerical studies using the Navier–Stokes equations (Knobloch and Moehlis 2000; Jiménez 2015).

In theory, the motion of two semi-infinite layers of ideal incompressible fluids of different densities is considered. The density of the lower liquid is higher than that of the upper, and surface tension is acting at the interface. In this case, the dispersion equation for infinitely small periodic perturbations is divided into two regions: a region of growth and a region of neutrally stable perturbations.

An important feature of the one-dimensional consideration is that the "boundary" modes are actually zero energy waves. There are two boundary modes with wave numbers K_1 and K_2. There is also a synchronism between them, $K_2 = 2K_1$. A resonance interaction occurs between these two modes. The case of three-dimensional geometry (two-dimensional interaction in the plane of the interface with mode structure along the vertical coordinate) has different features.

An important characteristic of boundary modes with zero energy waves is the time change of wave interactions. It is shown that nonlinear equations for slow-changed wave's amplitude describe wave–wave interaction in spatially homogeneous case, and yield self-similar solutions of "burst" type. In the solutions, the wave's amplitudes grow proportional to $A \sim (t - t_0)^{-2}$, where t_0 is the time of "burst." It is significant that in the case of near-threshold bursts, the "burst" time is inversely proportional to the square root of the small parameter (and not to the first power as in the case of ordinary bursts). The resonance interaction of internal waves in a stratified shear flow with discontinuous velocity and density profiles (in the case when these waves are boundary modes) has the same characteristics as the interaction of the boundary modes in a flow with discontinuous velocity and density profiles discussed above. This is the behavior of regular signals of finite amplitude interacting with the medium at the stability threshold. These are cases of signal amplification under the assumption that there is no well-developed turbulence.

It is well known, however, that turbulence can amplify large-scale motions. Therefore, the analysis of large-scale internal wave instability, a shear flow with parameters close to threshold for onset of turbulence, is an important case. Turbulent wave instability due to turbulent fluxes of momentum and buoyancy is seen. With the increase in wave energy, the corresponding variable component of the turbulent energy density also increases. A further analysis shows that the instability condition is very moderate: in essence, the only requirement is that the characteristic lengths of variation of the wave perturbation along the direction of stratification and those perpendicular to it be comparable. The process has a threshold character and sets in for $A > A_n$, where A_n depends on the local Richardson number and mode structure of the internal wave. The characteristic time of development of the instability is $\tau > 1/\Omega$, where $\Omega = K(C_f - U)$, K is the wave number of the perturbation perpendicular to the stratification, C_f is the phase velocity, and U is the unperturbed flow velocity. Because of the effect of turbulent fluxes on the wave, a small change in its amplitude over time occurs. The growth time of turbulent wave instability cannot be very close to $1/\Omega$.

Another type of instability, called secondary dissipative instability (Moiseev and Sagdeev 1986; Herbert 1988), is an interesting possible mechanism of direct generation of low-frequency hydrodynamic instabilities. Secondary unstable processes lead to spontaneous breakdown of the symmetry in dynamical systems. For example, the secondary instability "transfers" the initial one-dimensional process to two-dimensional process. Thus, the initial stage of the convective process in the ocean–atmosphere interface leads to the generation of cells with comparable vertical and horizontal dimensions and to simple topological structure. However, the field of turbulent fluctuations (generated by the convection itself or existing for some other reason) becomes gyrotropic. When such turbulence in the ocean–atmosphere interface is taken into consideration, a second stage of the convective process appears—the generation of large-scale structures with a horizontal scale considerably larger than the vertical, and with special stream lines.

The paper by Moiseev and Sagdeev (1986) attempts to briefly illustrate that "spontaneous regular behavior in a complex system, unrelated to the effect of external organizing fields, is the result of the development of a certain type of instabilities in this system." Characteristic features of chaos formation and structures depend significantly on the uniqueness of the secondary instabilities. Thus, owing to nonlinear interactions between wind waves and highly dynamic and localized turbulent flow environment (which could be a wake), there may occur two-dimensional coherent-like structures in the field of ocean surface roughness. This complex hydrodynamic event can be detected by high-resolution passive microwave radiometer-imager as distinct signatures of variable geometry (Chapter 6).

As a whole, instability-induced amplification mechanism may trigger multiple excitation of the wave number spectrum at high-frequency spatial intervals. The possible causes are associated with the following processes:

- Kinematics of long–short surface waves and wave–wave resonance interactions
- Surface wave–current interactions
- Oscillations of ocean boundary-layer parameters (wind speed, drag coefficient, roughness coefficient)
- Acoustics action and/or underwater sound effects ("parametric excitation")

Surface manifestations of these phenomena or joint effects can also be registered by a sensitive microwave radiometer.

2.4.7 Interaction between Surface and Internal Waves and Manifestations

The manifestation of the interaction between surface and internal waves is an important task of remote sensing investigations. Pioneer works (Hughes and Grant 1978; Hughes 1978) describe modulations of the surface wave by currents induced by internal waves. Usually, a special case of interaction between high-frequency surface waves and low-frequency internal waves (the surface waves are shorter than internal waves) is considered. This interaction appears strongest in the region of triple-wave resonance (Miropol'sky 2001), leading to the generation of intense surface waves (Phillips 1980), and to the surface wave "blocking" effect in the presence of internal waves (Basovich and Talanov 1977; Basovich 1979; Bakhanov and Ostrovsky 2002). Four-wave interactions come into effect in parameter regions where triple-wave processes are prohibited. They lead first to a modulation instability of surface waves due to their self-effect (Zakharov 1968), and to an additional modulation instability produced by the interaction between surface waves and an internal wave. This modulation instability has been considered for the case of a two-layer model of stratified fluid (Petrov 1979a,b), where the internal wave occurs at the interface between heavy and light fluids. Thus, in the case of discrete stratification, the internal wave turns out to be a potential one in contrast to the case of a continuously stratified medium, in which the internal wave has a *vortical character.*

The important step in the development of the hydrodynamic theory is the experimental testing of different schemes of surface–internal wave interactions. Well known radar signatures of coastal internal waves (Gasparovic at al. 1988) and models describing modulations of the surface wave by currents induced by internal wave (Hughes 1978).

For the first time, the effect of *five-wave* interactions between surface and internal waves was manifested and investigated experimentally using both radar and optical data (Mityagina et al. 1991). The airborne observations were conducted on a shelf of the Kamchatka Peninsula in the 1980s. Brightness modulations in the radar and optical images clearly reflect the presence of an intensive packet of internal waves. The conditions in the test area were: wind speed of 7 m/s and thermocline depth of about 27 m.

The primary analysis of the images shows that a three-wave interaction between two surface (s) and one internal (i) waves at low-frequency interval is observed:

$$\vec{K}_{s1} - \vec{K}_{s2} = \vec{K}_i, \quad \omega(\vec{K}_{s1}) - \omega(\vec{K}_{s2}) = \omega(\vec{K}_i). \tag{2.24}$$

(Four-wave resonance interaction involving three surface waves and one internal wave is prohibited by theory.)

To investigate surface–internal wave interactions in more detail, a digital Fourier processing of radar and optical data selected for the same ocean area was applied. As a result, an interpretation of the imaging data was made using the following resonance scheme of five-wave interaction:

$$2\vec{K}_{s0} = (\vec{K}_{s0} + \vec{K}_{s1}) + (\vec{K}_{s0} + \vec{K}_{s2}) + \vec{K}_i,$$

$$2\omega(\vec{K}_{s0}) = \omega(\vec{K}_{s0} + \vec{K}_{s1}) + \omega(\vec{K}_{s0} + \vec{K}_{s2}) + \omega(\vec{K}_i). \tag{2.25}$$

In this case, four surface wave (s) and one internal wave (i) components are considered: two components with identical vectors $\vec{K}_{s0}$ corresponding to the central maximum of wave number spectrum; two side components with vectors $\vec{K}_{s0} + \vec{K}_{s1}$ and $\vec{K}_{s0} + \vec{K}_{s2}$, and one internal wave with the vector $\vec{K}_i$. Conditions (2.27) were represented as $\Lambda_m^2 = \Lambda_s \Lambda_i$, where Λ_i is the wavelength of the internal wave, Λ_m is the wavelength of the surface modulation, and Λ_s is the wavelength of the initial surface wave ($\Lambda_i = 400$ m; $\Lambda_s = 42$ m; $\Lambda_m = 130$ m). A more detailed hydrodynamic theory, which describes a five-wave interaction and gives the values of coefficients of interaction, characteristics of modulation instability, and increments of surface waves, has not been developed.

It is important to note that the objective was to obtain the "ideal" oceanic conditions when a single packet of quasi-linear stationary internal waves was generated during a long period of time, and the wind speed was not changed. Another situation can be observed in the case of an intensive nonlinear internal wave interaction in a nonstationary wind field, or strong stratification of ocean–atmosphere boundary layer. The radar signatures of internal waves under these conditions have a complex spatial-nonuniform structure, and identification of any wave–wave synchronism on the images is difficult.

Important data were obtained during the 1992 Joint U.S./Russia Internal Wave Remote Sensing Experiment (Section 5.5.2). Strong interactions between gravity surface waves and packets of internal waves were investigated using airborne radar and optical techniques. After digital analysis of a large set of optical and radar data, it was found that the Fourier spectrum of the optical images has a clear multimode structure, that is, a multitude of separated spectral component (Etkin et al. 1995). The region of measured wave numbers corresponds to surface wind waves with wavelengths of $\Lambda_s = 5$–30 m, with a dominant wave component of $\Lambda_{sm} = 11$–13 m. Distinct spectral features manifested can be associated with strong nonlinear multiwave interactions, which rise due to complex modulations induced by the nonuniformity of surface currents.

In the ideal case of resonance multiple wave interactions, synchronism between n-wave components must satisfy the conditions:

$$\sum |\vec{K}_n| = 0 \quad \text{or} \quad \sum \left(|\vec{K}_n| - |\Delta\vec{K}_n| \right) = 0, \quad \sum \omega\left(|\vec{K}_n| \right) = 0, \qquad (2.26)$$

where $|\Delta\vec{K}_n|$ is the wave number mismatch for the interacting spectral components. However, it seems unlikely that a theory of multiple interactions (if any) will be suitable for the description of real-world wave phenomena. Deterministic hydrodynamic equations do not predict the behavior of dynamic systems with a large number of resonance interaction modes. Meanwhile, the analytic theory (Krasitskii and Kozhelupova 1995) defines weakly nonlinear resonance interactions between five trains of surface gravity waves.

We believe that multiple nonlinear quasi-resonance wave interaction processes can be described statistically. For example, the valuable wave vector components extracted from ocean optical or radar data should satisfy to certain spatial rules associated with a number of relevant physics-based interaction diagrams. Their statistical characterization can be made using multiplicative (or multifractal) cascade models, which may fit experimental diagrams. If it is possible at all, the main type of the interaction process could be specified and investigated. This research can be made using remote sensing observations only.

2.4.8 The Model of an Arbitrarily Stratified Ocean

The model of the interaction between surface and internal waves in a continuously stratified ocean of finite depth, with a triple-wave interaction process was suggested in the work (Rutkevich et al. 1989). Using the Euler equation of motion of incompressible fluid in a gravity force field, the basic evolutionary parabolic equation was obtained. The effect of the internal wave in this case occurs in the third-order of the perturbation method. The new dispersion relation for surface perturbations was deduced when the effect of modulation of a high-frequency surface wave by an internal wave was taken into account. The greatest contribution to this type of interaction is given by an internal wave, whose frequency is in resonance with the frequency of modulation of the surface wave. The dispersion relation was obtained from a parabolic nonlinear equation, which corresponds to the four-wave interaction process (one internal wave, two surface waves, and its envelope). Thus, the decay instabilities and modulation instabilities appear. These instabilities were also investigated, and different criteria were found. In the case of deep water, the instability has the most clearly defined drift character; in the case of shallow water, the instability has the largest increment, but it is realized in a narrow region of frequencies. The modulation instability due to wave–wave resonance interaction takes place in the narrow region of the angles between directions of propagation of the surface wave and its envelope. The modulation instability is absolute and gives rise to the generation of short wave packets. It can also stimulate the appearance of wave collapse or stationary surface soliton with the same frequency as that of the internal wave (Moiseev et al. 1999).

2.4.9 Model of Two-Layer Stratification and Interactions

The model (Petrov 1979a) describes the features of nonlinear interaction between surface and internal waves when the wavelength of surface waves is less than the wavelength of internal waves. The model of two-layer liquid, which is treated as a stratification model of the ocean with a sharply pronounced pycnocline, is considered. Unlike the model of arbitrary stratified ocean, this model gives different types of high-frequency surface nonlinear solitons, which are "blocked" by orbital currents due to internal waves.

This phenomenon is an example of self-action where the internal wave field has an external source. Thus, the effect "self-blocked" is apparent. To investigate nonlinear wave–wave interactions, the Hamiltonian variable principle is applied. The wave interaction Hamiltonian is presented in the form $H = H_s + H_i + H_{is}$, where H_s is the Hamiltonian of the interactions of the surface waves with one another (four-wave interaction); H_i is the Hamiltonian of the interaction of the internal waves with one another, and H_{is} is the Hamiltonian of the interaction between the surface and internal waves (three-wave interaction). As a result, analytic expressions are derived for the basic characteristics of solitary surface waves. The change of frequency of a short surface wave in the field of internal waves occurs due to two mechanisms: frequency variation due to modulation of the parameters of the medium and a Doppler frequency shift due to entrainment of the short wave by the moving medium. A concrete feature of the interaction studied in the model is the dominance of the Doppler mechanism changing the frequency of the surface waves by the moving fluid flow induced by the long internal wave near the free surface. Here, the parametric effect of frequency correction associated with the deformation of the free surface is found to be negligibly small.

The results of the theory are applied to investigate the behavior of surface–wave "envelope solitons," when they arise due to an interaction between surface waves and the pycnocline oscillations. In this case, the effect of modulation instability when the modulation grows due to nonlinearity of the surface wave and nonlinear oscillations of the interface is manifested. The length of the envelope solitons depends on a speed propagation of waves. Numerical estimations yield the values: $L = 400$ m and $A = 1 - 10$ cm, where L and A are the length and amplitude of the envelop solitons. Such an envelop can propagate over long distances (up to 10 km). Laboratory and numerical investigations (Slunyaev et al. 2013) show that group wave packets (or solitary groups) forming the surface envelop are relatively stable and significantly faster than the linear waves and even the nonlinear Stokes waves.

The effect of nonlinear damping of long surface waves due to the influence of an internal wave was also analyzed (Petrov 1979b). Surface–internal wave interaction is considered in the random phase approximation for the case of a two-layer ocean model. The damping decrement of long coherent surface waves, propagating in the field of *isotropic* and *anisotropic* random internal waves, is found. Some numerical estimations show that for natural ocean conditions, in both cases, the value of the damping decrement is about 10^{-4} second^{-1}.

This damping does not depend on the length of surface waves, and the result applies to perturbations of arbitrary form. It is important that the damping of long surface waves due to their interaction with random internal waves can reach the limiting maximum values for the theory of a weak nonlinear interaction. In this case, the surface wave will be damped by *a factor of e* over a distance of ~1000 km from the moment it starts the surface–internal wave interaction.

Summarizing, we list the following aspects of wave hydrodynamic theory that present a great interest for remote sensing studies:

- Nonlinear interactions of multiscale surface waves
- Generation of 2D and 3D surface wave structures
- Modulation of short surface waves by long surface waves
- Generation and evolution of the surface waves induced by nonuniform current field
- Damping of surface waves due to turbulence
- Development of surface wave instabilities and effects of wave number spectrum excitation
- Nonlinear dynamics and spatiotemporal reorganization of subsurface hydrodynamic fields, including shear flows
- Development of thermohaline convective processes in subsurface ocean layer (Section 2.6)

A sketch of presented theories demonstrate an important role of *large-scale surface hydrodynamic processes* (disturbances) generated by nonlinear wave–wave interactions in stratified ocean. These results have a principle value in ocean microwave remote sensing and advanced applications.

2.5 Wave Breaking and Disperse Media

Wave breaking is one of the most abundant nonlinear phenomena in the ocean. As a result of wave breaking and intensive mixing of air and water, various types of two-phase disperse media—bubbles, foam, whitecap, spray, aerosol, and their aggregations—occur at the ocean–atmosphere interface. This phase transition leads to considerable changes in microwave emission characteristics.

In this chapter, we discuss these fascinating events in context with microwave remote sensing. Several books (Bortkovskii 1987; Kraus and Businger 1994; Massel 2007; Sharkov 2007; Steele et al. 2010; Toba and Mitsuyasu 2010; Babanin 2011; Soloviev and Lukas 2014) provide more detailed information concerning oceanographic observations and analysis of wave breaking fields.

2.5.1 Wave Breaking Mechanisms

The criteria for individual wave breaking event in deep water have been formulated first by Stokes in 1847 and later by Michell (1893). They are the following (Massel 2007):

- The particle velocity of fluid at the wave crest equals the phase velocity

- The crest of the wave attains a sharp point with an angle of 120°
- The ratio of wave height to wavelength is approximately 1/7
- Particle acceleration at the crest of the wave equals 0.5 g

In real-world wave breaking process is conditioned by a disruption of the equilibrium between the redistribution of energy into the wave spectrum on the one hand, and by atmospheric excitation (pumping) of wind waves in the range of the spectral maximum on the other hand. As a result of this redistribution, which occurs very slowly, wind-generated waves become unstable and break. There is a number of dynamic models and numerical simulations of wave breaking process (Hasselman 1974; Melville 1994; Terray et al. 1996; Chen et al. 1997; Makin and Kudryavtsev 1999; Banner and Morison 2010; Irisov and Voronovich 2011; Chalikov and Babanin 2012). Most of them are based on the statement that the wave spectrum is changed in low-frequency interval, causing considerable dissipation of wave energy. The spectrum variations are described by the balance kinetic Equation 2.12. In principle, spectral-based models and solutions become applicable for microwave remote sensing in the case when the impact of wave breaking on radar backscatter or emissivity is considered at large averaging.

For statistical characterization of wave breaking, Phillips (1985) introduced the so-called multiscale breaking rate $\Lambda(c)dc$, which is the averaged length of the breaking crest per unit area traveling at velocities in the range $(c,c + \Delta c)$. Phillips's theoretical concept is used for estimating the total energy dissipation rate due to wave breaking

$$E = b\rho g^{-1}\int c^5\Lambda(c)dc, \qquad (2.27)$$

and the momentum flux from the waves to currents

$$M = b\rho g^{-1}\int c^4\Lambda(c)dc, \qquad (2.28)$$

and also active whitecap fraction

$$W_A = T_{phil}\int c\Lambda(c)dc, \qquad (2.29)$$

where g is gravity, ρ is the density of water, and b is a numerical constant or "breaking parameter"; T_{phil} is an average bubble persistence time introduced by Phillips (1985).

Remote sensing measurements (Phillips et al. 2001; Melville and Matusov 2002; Thomson and Jessup 2009; Callaghan et al. 2012; Gemmrich et al. 2013) demonstrate robust relationship between $\Lambda(c)$, E, and M and wave breaking statistics. In particular, Melville and Matusov (2002) found that the momentum flux from the waves to currents and wave dissipation are proportional to wind speed $M,E \sim V^3$ and dominated by intermediate scale waves. The Phillips's $\Lambda(c)$-based model is developed for remote sensing applications as well (Reul and Chapron 2003; Irisov 2014; Irisov and Plant 2016). As a whole, single-parameter whitecap coverage models have certain limitations (Guan et al. 2007) because they do not describe a variety of foam/whitecap properties.

Another mechanism affecting the process involves the effects of fluctuations of air flow over the ocean surface. The intensity of wave breaking depends on the presence of the surface wind drift and swell (Phillips and Banner 1974). In this model, the limiting wave height is less, and is estimated to be approximately 1/3 from the Stokes wave limiting configuration.

In the open ocean, the wave breaking process begins earlier than is suggested by both theories. In nature, the influence of surface current and wind speed fluctuations are significant. As a result, conditions of large- and small-scale surface wave–wave interactions are changed, and the limiting wave configurations are determined by dynamical parameters of the ocean boundary layer (Kitaigorodskii 1984).

The geometry, structure, and evolution of breaking wave profiles is illustrated schematically in Figure 2.5 (Bunner and Peregrine 1993). In the wave breaking zone, two-phase turbulent flow is formed analogously to the flow on a downhill surface. Although hydrodynamic theory of wave breaking is not developed fully, some theoretical estimates give good agreement with laboratory measurements at the early stages of the wave breaking process (Longuet-Higgins and Turner 1974; Rapp and Melville 1990).

In nature, the following types of wave breaking are distinguished:

- Spilling breakers—wave crests spill forward, creating foam and turbulent water flow

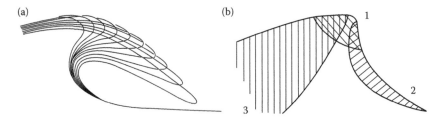

FIGURE 2.5
Temporary evolution of wave profile (a) and dynamical zones of wave breaking (b). (1) Speed of water particles is more than phase speed of a surface wave; (2) acceleration of water particles is more than the acceleration of the gravity g; (3) acceleration of the water particles is less then g/3. (Adapted from Bunner, M. L. and Peregrine, D. H. 1993. *Annual Review of Fluid Mechanics*, 25:373–397; Cherny I. V. and Raizer V. Yu. *Passive Microwave Remote Sensing of Oceans*. 195 p. 1998. Copyright Wiley-VCH Verlag GmbH & Co. KGaA. Reproduced with permission.)

- Plunging breakers—wave crests form spectacular open curl; crests fall forward with considerable force
- Collapsing breakers—wave fronts form steep faces that collapse as waves move forward
- Surging breakers—long, relatively low waves whose front faces and crests remain relatively unbroken as waves slide up and down

The most adequate quantitative study of wave breaking is based on the numerical Navier–Stokes simulations (Lin and Liu 1998; Chen et al. 1999; Iafrati 2009; Ma 2010; Higuera et al. 2013; Lubin and Glockner 2015). For this goal, methods and algorithms of computational fluid dynamics are used. Numerical experiments yield impressive and probably the most valuable results concerning generation, propagation, interaction, and evolution of one-, two-, and even three-dimensional nonlinear surface waves, including wave breaking shape configurations and patterns. However, the applications of Navier–Stokes solutions and simulations for real-world ocean environment is still a difficult task due to stochastic and multiscale nature of wave motions.

Wave breaking is the main process causing the phase transition in the ocean–atmosphere system. As a result, two-phase disperse media—bubbles, spray, foam, and whitecap—are generated on the ocean surface that is important in air–sea interactions and mass transfer (Bortkovskii 1987; Melville 1996).

The classification of oceanic disperse media are shown in Table 2.2. Their microstructure varies dramatically; it is difficult to predict the behavior parameters using conventional hydrodynamic theory except a common-sense statement that it is a heterogeneous mixture of air and water. A more

TABLE 2.2

Classification and Parameters of Ocean Disperse Media

Main Properties	Whitecap (Plume)	Foam Streaks	Spray	Aerosol	Subsurface Bubbles
Area coverage (m)	0.5–10	3–30	10–20 (local clouds)	>1000	>1000
Averaged thickness of layer (m)	0.01–1.0 multiple	0.01–0.01 monolayer	0.2–1.5 > 1.5	0.5–10	0.01–0.05
Volume water concentration (%)	20–50	<5–10	0.01–0.1	<0.01	0.5–1.0
Size of particles (cm)	0.5–1.0	0.01–0.5	0.01–0.1	<0.01	<0.01
Lifetime stability	Seconds (unstable)	Minute, hours (stable)	Seconds (unstable)	Minutes (stable)	Hours (stable)

Source: Cherny I. V. and Raizer V. Yu. *Passive Microwave Remote Sensing of Oceans.* 195 p. 1998. Copyright Wiley-VCH Verlag GmbH & Co. KGaA. Reproduced with permission.

FIGURE 2.6
Main types of oceanic disperse media with different microwave properties. (a) Hydrodynamic plume. (b) Two-phase turbulent flow. (c) Dense sea spray. (Adapted from (a) http://www.wallpapersxl.com/wallpaper/1680x1050/syndicate-wave-breaking-the-free-information-society-208959.html; (c) http://hqworld.net/gallery/details.php?image_id=5518&sessionid=3233f4f17412bcf40f60b1358542959f.)

complicated for analysis and, perhaps, more realistic definition of oceanic disperse media should be based on detailed nature observations and measurements. Indeed, an entire foam/whitecap/spray system can be divided into several types as shown in Figure 2.6. They exhibit highly variable physical and electromagnetic properties. The most dynamic disperse object (Figure 2.6b) can be defined and adequately modeled as a two-phase turbulent composition stratified flow of gaseous and liquid particles having different geometry, size, distribution, and aggregative stability. Such a description is more realistic in view of advanced remote sensing studies than just statistical or matrix air–water mixtures.

2.5.2 Foam and Whitecap

Foam belongs to the class of colloidal systems that includes two phases: gas and liquid. The physical state of foam is defined by its stability and by inside disperse structure. Therefore, all colloids are considered heterogeneous systems with a great area of the air–water interface, the foam is basically unstable. Foam that exists for a few seconds may be considered unstable, but foam that exists for some minutes or hours may be considered stable.

The structural classification of foam as a disperse colloid system is as follows (Bikerman 1973; Weaire and Hutzler 1999; Zitha et al. 2000; Breward 1999):

- Mono- or polydispersed system of the ideally spherical particles (gaseous bubbles), chaotically distributed in the liquid medium
- Continuous structure of close-packed spherical-like bubbles
- Cellular system of close-packed bubbles of irregular polyhedral shapes
- Dry foam consisting of thin liquid films which are formed in polyhedral cells

There are two categories of sea foam: dynamical foam (its lifetime is less than 1 min) and stable foam (its lifetime is more than 1 min). The following terminology is used for describing the physical conditions of sea foam: "white water," "whitecaps," "thin foam," "foam streaks," and "foam patches." These categories are conventional and are based on marine observations. The two simple terms "foam" and "whitecaps" are used in remote sensing in order to distinguish their microwave emission characteristics, although this terminology may not fully describe a variety of environmental two-phase structures and situations.

As shown by detailed nature investigations (Miyake and Abe 1948; Abe 1963; Raizer and Sharkov 1980; Bortkovskii 1987), the microstructure of thin stable foam patches represents a concentrated gas emulsion of closely packed bubbles, or the so-called emulsion monolayer, located on the sea surface. Possible bubble diameters into the monolayer are 0.01–0.5 cm. The other type represents the so-called polyhedral cell foam of thickness ~1.0–2.0 cm. The size of individual cells can reach several centimeters. The dynamic properties of these two structures are different: foam monolayer is stable, but polyhedral cell foam is unstable. Laboratory foam samples are shown in Figure 2.7 (Raizer and Cherny 1994).

In the real world, the most abundant unstable disperse structure is the so-called whitecap bubble plume (which should not be confused with deep ocean bubble plumes in acoustics). We define the plume as a violent, extremely saturated, two-phase turbulent flow separated over a wind-wave crest. Plumes occur due to a cascade collapse of the massive crests of large-scale gravity waves that occur at very high winds. Intensive air–water mixing and the gravity force produce free-falling jets consisting of conglomerates of air bubbles and water particles of complicated structure and geometry. The size of the particles may vary from 0.1 to 10 cm, depending on the aeration process. The impact of the massive plume on microwave emission is strong enough even at S and L bands (Raizer 2008).

Whitecap phenomenon has been explored in open oceans by a number of oceanographers (Blanchard 1971; Monahan and MacNiocaill 1986; Bortkovskii 1987; Lamarre and Melville 1994), but little is known about the physical properties and disperse microstructure inside a whitecap plume. The photograph in Figure 2.6a illustrates an example: powerful whitecap plumes formed during wind wave breaking. The effective thickness of dense whitecap plume can vary from several dozens of centimeters to several meters. Indeed, laboratory studies (Monahan and Zietlow 1969; Zheng et al. 1983; Peltzer and Griffin 1988; Callaghan et al. 2012, 2013) demonstrate the existence of the relationship between microstructure parameters and aggregate stability of thick layers of dynamic polydisperse foam and whitecap plume.

2.5.3 Wave Breaking and Foam/Whitecap Statistics

Wave breaking and foam/whitecap statistics are important attributes in many oceanographic and remote sensing studies. Wave breaking events

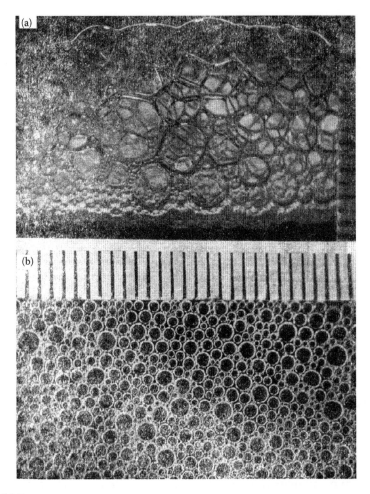

FIGURE 2.7
Foam microstructure. (a) Polyhedral foam structure. (b) Emulsion monolayer of bubbles on the water surface. (Raizer's original pictures. Adapted from Cherny, I. V. and Raizer, V. Y. 1998. *Passive Microwave Remote Sensing of Oceans*. Wiley.)

are clearly visible in good-resolution aerial photos (optical images) in the form of distinct geometrical objects representing whitecap and foam structures on the ocean surface. At high winds and strong gales, whitecap and foam area fractions vary significantly according to the Beaufort scale that can be recorded using aerial photography as well. The wave breaking activity (intensity) is not only defined by wind wave dynamics but also depends on parameters and the stability of the ocean–atmosphere boundary layer, subsurface wave processes, and surface currents. Therefore, wave breaking events can be observed sometimes at low winds.

Systematic airborne observations of the ocean surface using high-quality aerial photography and a number of passive microwave radiometers were conducted in the Pacific Ocean first from the Antonov An-30 aircraft (1981–1986) and then from the aircraft laboratory Tupolev Tu-134 SKh with the registration number CCCP-65917 (1987–1992) equipped with the six-band optical MKF-6 aerial photo camera (made by Carl Zeiss, Germany), the Ku band (2.25 cm wavelength) side-looking airborne radar (SLAR "Nit"), and several passive microwave radiometers at wavelengths 0.8, 1.5, 8.0, and 18 cm (Section 5.5.2).

In particular, observations of large-scale surface dynamics and wave breaking fields from an aircraft at altitudes from 300 to 5000 m have provided high-quality optical imagery of the ocean surface at variable winds and fetch conditions (Raizer et al. 1990; Raizer 1994). For example, at a flight altitude of 5000 m, the size of the MKF-6 single frame is ~3 km × 5 km with a spatial resolution of ~3–5 m. The required information is obtained using digital image processing, including two-dimensional fast Fourier transform (FFT) and morphological and statistical analysis of foam/whitecaps objects registered in the optical images.

At that time, for the quantitative analysis of foam/whitecaps coverage and geometrical statistics, the following metrics have been introduced: area A, perimeter P, maximum and minimum linear size L_{max} and L_{min} and nondimensional topological metrics P^2/A and $A/L_{max}L_{min}$ for each single foam and/or whitecap object visible in the image. As a result of the digital processing of a huge body of optical data, statistical distributions of foam/whitecap metrics were defined for different ocean surface conditions. For example, strong transformations of foam and whitecaps coverage statistics (metrics) are observed at limited fetch conditions (Raizer 1994).

Moreover, the dependences of averaged total foam + whitecaps area fraction W, % on wind speed U were measured from airborne optical data with a high degree of accuracy.

We have to note that a large number of optical and video observations have been made in order to define the dependency W(U) in the open ocean (Monahan 1971; Ross and Cardone 1974; Monahan and O'Muircheartaigh 1980, 1986; Bortkovskii 1987; Wu 1988b; Monahan and Lu 1990; Bortkovskii and Novak 1993; Zhao and Toba 2001; Stramska and Petelski 2003; Lafon et al. 2004; Bondur and Sharkov 1982; Sugihara et al. 2007; Callaghan et al. 2008; Callaghan and White 2009).

However, there is still some uncertainty in data collections and empirical approximations of the function W(U). The problem is not only in a great variety of ocean environmental conditions. Indeed, foam and whitecap objects visible in the optical images have different contrasts, blurring boundaries, and unsharp contours. This circumstance makes it difficult to accurately measure foam and whitecap objects, that leads to errors in computing their geometric parameters and area fractions. A special algorithm for automatic recognition and analysis of foam/whitecap objects in the optical images has been develop and applied (Raizer and Novikov 1990) in order to obtain statistically reliable data and dependencies.

TABLE 2.3

Empirical Coefficients of the Power-Law Formula for Whitecap Coverage versus Wind Speed W = aV^b

Author	a ($\times 10^{-6}$)	b
Blanchard (1963)	440	2.0
Monahan (1969)	12	3.3
Monahan (1971)	13.5	3.4
Tang (1974)	7.75	3.23
Wu (1979)	1.7	3.75
Monahan and O'Muircheartaigh (1980)	3.84	3.41
Wu (1988b)	2.0	3.75
Hanson and Phillips (1999)	0.204	3.61

In general, the power-type formula

$$W = aU^b \qquad (2.30)$$

is used in microwave remote sensing applications. Here, W represents the instantaneous fraction of the sea surface covered by foam and whitecap, U is the 10-m-elevation wind speed, and a and b are empirical constants. For example, an optimization of data sets (Monahan and O'Muircheartaigh 1980) yields the values $a = 3.84 \cdot 10^{-6}$ and $b = 3.41$ in the range of wind speed $5 < U < 25$ m/s. Table 2.3 and Figure 2.8 summarize some proposed approximations (Zhao and Toba 2001).

During the last several years, satellite-based microwave observations have been used for the assessments of the global distribution of foam and whitecap coverage in the world's oceans (Anguelova and Webster 2006; Anguelova et al. 2009; Bobak et al. 2011; Salisbury et al. 2013; Albert et al. 2015; Paget et al. 2015). The algorithm is based on the retrieval of wind vector and utilization of microwave data using empirical relations (2.30). Theoretically, the value of foam and whitecap area fraction, estimated by optical and microwave data should not be the same (due to the difference in physical mechanisms of ocean microwave and visible radiances); however, the method gives a promising statistical result. The best option is still to provide combined optical–microwave observations in order to improve the information performance and reduce possible errors.

It is important to note that at fully developed storm conditions, the effect of "W-saturation" can take place. The saturation is not described by simple power approximations. In this case, area fraction W does not depend on the ocean surface state significantly, and the behavior of function W(U) is defined through the energetic balance in the wind–wave system.

A novel approach for providing statistical characterization of foam/whitecap activity is based on the fractal (or multifractal) dimension

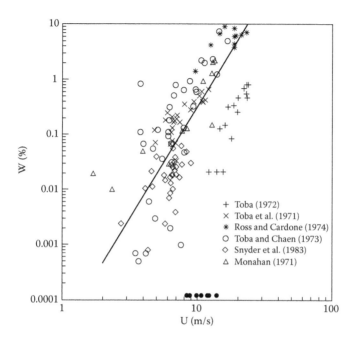

FIGURE 2.8
Empirical dependencies of whitecap area fraction on wind speed. The solid line is expressed as $W = 2.98 \times 10^{-5} U^{4.04}$. (Adapted from Zhao, D. and Toba, Y. 2001. *Journal of Oceanography*, 57(5):603–616.)

formalism. Fractal geometry describes spatial self-similarity and scaling of dynamic systems and natural objects (Mandelbrot 1983). The application of fractal analysis for exploring wave breaking fields by remotely sensed (infrared, optical, video) data is actually a challenging task. Nevertheless, such studies based on airborne optical observations have been conducted earlier (Raizer and Novikov 1990; Raizer et al. 1994; Sharkov 2007).

Let us consider some principle results. There are two basic procedures to compute the fractal dimension of wave breaking fields by optical images. The first procedure is based on the so-called box counting method, which yields the Hausdorff dimension

$$D_H = \lim_{r \to 0} \frac{\log N(r)}{\log(1/r)}, \qquad (2.31)$$

where $N(r)$ is the smallest number of squares with the side r required to completely cover the data set (binary image, for example).

The second procedure was based on a simple relationship between area A and perimeter P of a single fractal-like geophysical object. In our case, the fractal dimension of each individual object (foam streak or whitecap)

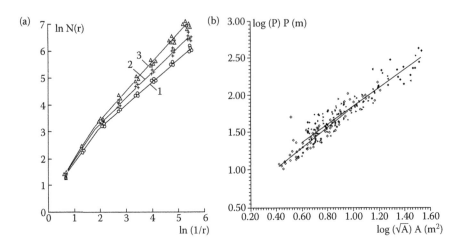

FIGURE 2.9

Fractal characteristics of foam and whitecap coverage obtained from optical data. (a) ln N(r) vs. ln(1/r) for total foam and whitecap coverage. The value of fractal dimension corresponds approximately to three gradations of the Beaufort wind force: $1 - 3 \div 4$ ($D_H = 1.05$); $2 - 4 \div 5$ ($D_H = 1.15$); and $3 - 5 \div 6$ ($D_H = 1.25$). (b) log-log plots of perimeter (P) as a function of square root from area ($\sqrt{A}$) for foam streaks (*) and whitecaps ($\lozenge$). Sold lines are linear least-square fits in different ranges of area A. Beaufort wind force is 4. (Adapted from Cherny, I. V. and Raizer, V. Y. 1998. *Passive Microwave Remote Sensing of Oceans*. Wiley.)

visible in optical image is estimated from the so-called area–perimeter relationship

$$P \sim (\sqrt{A})^{D_s}, \tag{2.32}$$

where A is the area and P is the perimeter of a selected individual foam or whitecap object. The averaging of fractal dimension D_s at an object's ensemble on the image yielded a mean value of fractal dimension $\bar{D}_S$.

The results of the fractal analysis of a large set of ocean optical data are shown in Figure 2.9. For a moderately stormy ocean, it was found that the Hausdorff fractal dimension changes to within $D_H = 1.1$–1.3 (Figure 2.9a). At the same time, the regression coefficients of the area–perimeter relationship (2.32) gives the value of the fractal dimension $\bar{D}_S = 1.39$ and 1.23 for the whitecaps and foam streaks, respectively (Figure 2.9b). Statistically, the computed fractal dimensions differ for foam streaks (patches) and whitecaps.

An important aspect is dynamics of wave breaking in the field of ocean internal waves. It is a well-known fact that the influence of internal waves causes the wave breaking wave statistics to change. Both the frequency of wave breaking acts (intensity) and the total area fraction of foam/whitecap coverage increase due to interactions between internal and surface waves and the induced surface current. The wave breaking intensity and foam/whitecap geometry can be associated with the structure of the internal wave field.

The aircraft experiments conducted in the North Pacific Ocean during the period 1981–1991 show that the fractal dimension measured by optical signatures of foam and whitecap varies in the presence or absence of an internal wave source. This effect is manifested at moderate as well as high winds. It can be explained by the change of wave breaking intensity (frequency of braking acts) and statistics of foam and whitecap coverage.

For example, under the influence of the surface current, stochastic wave breaking structures may turn into orderly type patterns. Additionally, the surface current gradients effect the speed of energy dissipation that leads to the change of the regime of wave breaking process. Supposedly, the intensity (frequency) of wave breaking increases in the zone of the current's convergence, and decreases in the zone of the current's divergence. Internal waves cause the spatial modulation of surface waves and thus accelerate the wave breaking process even at low winds.

Because there are some difficulties in adequate hydrodynamic and statistical modeling of real-world wave breaking processes, the required information can be obtained using high-resolution optical–microwave observations. Remotely sensed data allow researchers to employ different digital methods and techniques for analyses of spatially statistical characteristics of wave breaking and foam/whitecap events. For example, the fractal dimension of foam/whitecap coverage computed by optical–microwave imaging data can provide a quantitative criterion of stormy sea state in addition to the Beaufort wind force scale.

2.5.4 Surface Bubble Populations

Surface bubble populations represent clusters of individual spherical bubbles floating on the sea surface. They are considered as an intervening type between a thin layer of sea foam and the near-surface bubble clouds at the upper ocean. The shape of the surface resembles an ensemble of hemispherical shells above the water surface. Bubble populations usually cover huge ocean spaces at high winds and may not be visible directly from a ship or an aircraft. Sometimes, the interference picture from the bubble's films is observed. However, bubbles produce considerable changes in ocean microwave emission at millimeter and centimeter range of the wavelengths.

The main environmental source of bubble populations is wave breaking and foam/whitecap decay. Mechanisms of generation and physical properties of surface bubbles have been studied by many authors over the years (Johnson and Cooke 1979; Johnson and Wangersky 1987; Walsh and Mulhearn 1987; Baldy 1988; Wu 1988a; Monahan and Lu 1990; Thorpe et al. 1992; Thorpe 1995; Bowyer 2001; Woolf 2001; Leifer et al. 2006). These and other investigations show that initial bubble populations generated after wave breaking are very dense and compact. The size distribution of surface bubbles is known for a wide range from 0.01 to 1 cm. Bubble stability and size depend on density, temperature, and salinity of seawater (Hwang et al. 1991; Slauenwhite and Johnson 1999; Wu 2000).

Bubbles are also generated due to mechanic mixing of air and water behind a moving body. In this case, two-dimensional bubble patterns—"bubble wake" and/or "bubble jet"—are produced. The ship's bubble wake may exist in the surface for a long time. The presence of bubble wake causes the change of propagated acoustic signals (Trevorrow et al. 1994; Phelps and Leighton 1998; Stanic et al. 2009).

Finally, bubble populations occur on the surface under the influence of different marine biological processes, organic particles, surface-active materials, and pollutants (Garrett 1967, 1968; Clift et al. 1978; Johnson and Wangersky 1987). Organic films stabilize the bubble lifetime, creating dense bubble patches at the air–sea interface. These bubbles are coated with surfactant material that provides a stabilizing mechanism against surface tension pressure and gas diffusion. Extended surface monolayers of stabilized bubbles produce a peculiar kind of environmental "electromagnetic diffracting screen or grating" that affects the propagation of acoustic signals.

2.5.5 Spray and Aerosol

Sea spray can be produced by the wind through various mechanisms: direct shearing of wave crests by wind, aerodynamic suction at the crests of capillary waves, and bursting of air bubbles at the water surface. Spray represents a system of liquid particles (droplets) located above the sea surface, whereas the near-surface aerosol comprises both liquid and solid sea salt particles. The main environmental source of spray and aerosol production is the injection of jet droplets due to wave breaking, foam/whitecap, and bubble bursting events (Figure 2.10). Spray and aerosol are important components of ocean–atmosphere system and make a major contribution in the air–sea exchange and fluxes.

The structure and dynamics of sea spray and the near-surface aerosol were investigated in nature and laboratory by many authors (Blanchard 1963, 1983, 1990; Monahan 1968; Wu 1979, 1989a, 1990a,b, 1992a,b, 1993; Monahan et al. 1982; Bortkovskii 1987; Andreas 1992; Fairall et al. 1994; Spiel 1994, 1998; Andreas et al. 1995, 2010; Anguelova et al. 1999; Lewis and Schwartz 2004; Kondratyev et al. 2005; Callaghan et al. 2012; Veron et al. 2012; Norris et al. 2013; Grythe et al. 2014; Veron 2015).

At least three main parameters have to be considered in context with microwave remote sensing of the ocean: (1) volume (bulk) concentration; (2) size distribution; and (3) height distribution of droplets/particles above the surface. These parameters can be incorporated to ocean microwave emission models (Chapter 3).

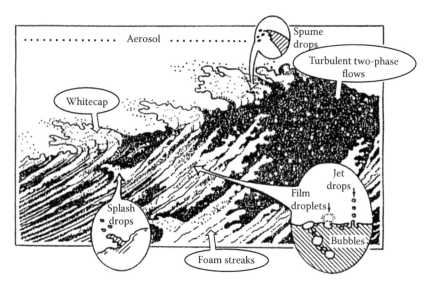

FIGURE 2.10
Oceanic disperse media. Spray production. (Adapted and updated from Andreas, E. L. et al. 1995. *Boundary-Layer Meteorology*, 72:3–52; Raizer, V. 2007. *IEEE Transactions on Geoscience and Remote Sensing*, 45(10):3138–3144. Doi: 10.1109/TGRS.2007.895981.)

Classical literature data concerning ocean spray characteristics are shown in Figure 2.11. It is known from field experiments that the size distribution of the ocean spray follows the power law $p \sim r^{-n}$, where r is the radius of a droplet. Exponent n changes from 2 to 8, depending on wind conditions. The range of the droplet's diameter is quite wide: 10^{-4}–10^{-2} cm. The height of a dense spray layer above the ocean surface is about 10–40 cm and depends significantly on the droplet-generating mechanism (Blanchard 1963; Bortkovskii 1987). The mass concentration of the water in the ocean spray near the surface is about 10^{-4}–10^{-1} g/cm^3. Dense layers of spray are located mostly around the breaking crest of wind waves. The vertical distributions of size and volume concentration of spray are highly nonuniform. Usually, small-sized droplets and aerosol cover the foam-free water surface; large-sized droplets are formed mostly over the foam and whitecaps areas.

2.5.6 Subsurface (Underwater) Bubbles

Bubbles in the subsurface ocean layer (<1 m) are important elements in ocean–atmosphere gas exchange. Bubble clouds generated by breaking waves create highly concentrated aeration layers that cause significant changes in the electromagnetic skin depth. Therefore, the near-surface bubbles can produce high-contrast signatures at microwave frequencies (Chapter 3).

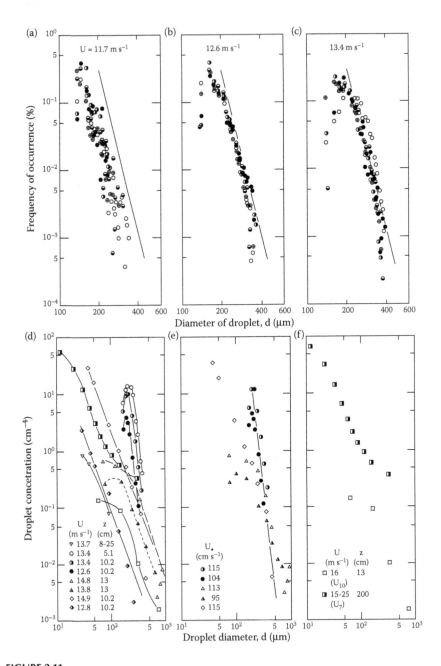

FIGURE 2.11

Sea spray size distribution at the near-surface ocean. The probability density of occurrence at different wind speed: (a) U = 11.7 m/s; (b) U = 12.6 m/s; (c) U = 13.4 m/s. (d)–(f) Droplet concentration for different surface conditions. U is the wind speed; μ_* is the friction velocity; z is the elevation above the mean water surface. (Adapted from Wu, J. 1979. *Journal of Geophysical Research*, 84(C4):1693–1704.)   (Continued)

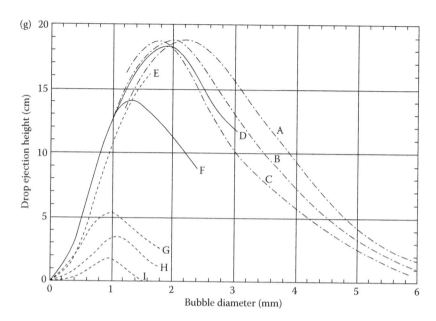

FIGURE 2.11 (*Continued*)

(g) Jet drop height as a function of bubble diameter, temperature, and salinity. A—4°C seawater; B—16°C seawater (top drop); C—30°C seawater (top drop); D—22–26°C seawater (top drop); E—4°C seawater (top drop); F—21°C distilled water (top drop); G—22–26°C seawater (2nd drop); H—22–26°C seawater (3rd drop); I—22–26°C seawater (4th drop). (Adapted from Blanchard, D. C. 1963. In *Progression Oceonography*, pp. 73–202. Pergamon Press. Doi: 10.1016/0079-6611(63)90004-1.)

Sources and mechanisms of bubble production in the ocean are the following:

- Formation of globally distributed surface bubble clouds due to ocean–atmosphere interactions
- Migration of methane and carbon dioxide gases from deep water to the surface
- Mechanical mixing of air and water due to wave breaking activity
- Impact of raindrops and spray on the surface
- Cavitation due to the rotation of blades of a ship's propeller (hydrodynamic cavitation)
- Propagation of intense sound wave into oceanic water (acoustic cavitation); cavitation currents induced by strong turbulent flow over a moving body
- Underwater explosions, earthquakes, nuclear bomb tests, submarine and torpedo destructions
- Very high-speed underwater vehicles

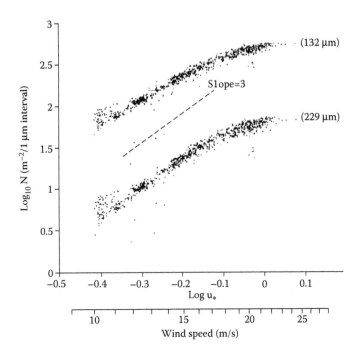

FIGURE 2.12
Inferred bubble population density at subsurface ocean layer depending on wind speed (U) and the friction velocity (μ_*). Underwater ambient noise measurements at acoustic frequencies 25.0 and 14.5 kHz. Resonance radii of bubbles: 132 and 229 µm. (Adapted from Farmer, D. M. and Lemone, D. D. 1984. *Journal of Physical Oceanography*, 14(11):1762–1778.)

Bubble properties in the upper ocean layer have been studied using acoustic sounding and measurements of the ambient noise (Kolobaev 1976; Kerman 1984; Vagle and Farmer 1992; Leighton 1994). Nature observations (Medwin 1977; Clift et al. 1978; Johnson and Cooke 1979; Mulhearn 1981; Thorpe 1984; Medwin and Breitz 1989; Wu 1989a; Wu 1992a; Anguelova and Huq 2012) show that the size distribution of subsurface bubbles is governed by the power law $p \sim a^{-n}$, where n is changed from 3.5 to 5.5 at the range of bubble's radius $a = 10^{-4}$–10^{-1} cm. Figure 2.12 illustrates bubble distributions measured by echo sounder (Farmer and Lemone 1984). The volume concentration of gaseous bubbles in the upper layer of the ocean can reach 20% and more. Figure 2.13 shows estimates of bubble volume concentrations for typical situations. These data demonstrate how bubbles production and concentration can depend on the type of the internal source.

2.5.7 Surface Films, Oil Slicks, and Emulsions

Surface films are encountered in the ocean very often. Along with their origin, they can be divided into two categories: natural and artificial. Natural

FIGURE 2.13
Bubble production in the ocean and gradations of the volume concentration (C). (a) Natural oxygen aeration, C < 0.1; (b) cavitating flow, 0.1 < C < 0.3; (c) deep-water bubble plume, C > 0.3. (Adapted and modified from (a) http://michaelprescott.typepad.com/.a/6a00d83451574c69e20 1b8d0890adb970c-pi.)

slicks are formed due to the chemical and biological processes in the ocean; they are known as surface-active films. Artificial slicks appear on the ocean surface as a result of the anthropogenic activity of human beings. These are polluting surface-active films of oil (or petroleum) and other synthetic and detergent oil products. Generally, they agglomerate in coastal economically advanced zones.

Surface-active films are manifested in the form of "smooth surface" areas. The thickness of such films is equal to several monomolecular layers (10^{-7}–10^{-6} cm). Organic surface-active films change the optical characteristics of surrounding water. Sometimes, their presence causes the appearance of anomalous phenomena in the reflected light due to variance of the slopes of short gravity waves. There are various configurations of the slicks: they can be long streaks, oriented along the direction of the wind, or separate areas reminding us of Langmuir's cells. Surface slicks are indicators of marine processes and internal waves (Kerry et al. 1984; Gade et al. 2013).

For remote sensing applications, the most interesting presents effects of interaction of surface films and wind waves. Monomolecular surface films strongly damp small-scale wave components, resulting in a variability of the wind wave spectrum (Hühnerfuss and Walter 1987; Alpers and Hühnerfuss 1989; Wu 1989b; Ermakov et al. 1992; Wei and Wu 1992; Gade et al. 2006; Ermakov 2010). The variation of the spectral density of the wave energy due to the damping effect is registered in a wide range of frequencies: from 3 to 15 Hz. Radar observations (Hühnerfuss et al. 1994; Espedal et al. 1996; da Silva et al. 1998; Karaev et al. 2008) also show that surface-active films sharply change regimes of the generation of gravity–capillary waves in the wind field that significantly changes the backscatter signal. The largest decrease in the L band brightness temperature was manifested due to the influence of a monomolecular oleyl alcohol film on the sea surface (Alpers et al. 1982). However, the authors explain this effect by anomalous dispersion of the film permittivity but not by damping surface wave components.

Unlike the surface-active films, the oil films never make monomolecular layers. The range of the thickness of typical oil films is 10^{-4}–1 cm. Thin films of crude oil give a silver glance; thicker oil films have a dark color without interference painting. Layers of water/oil emulsion remind the observer of a thick "chocolate mousse." The thickness of such emulsion layers can equal several centimeters.

Experimental studies (Creamer and Wright 1992; Tang and Wu 1992) show the oil film's effect on the process of wind-waves generation. On a wave crest, the thickness of the film is usually more than that of a wave hollow. Polluting films depress high-frequency components of surface wave spectrum stronger than organic films. Along with that, they brake the mass–heat exchange between the ocean and the atmosphere. In the slick area, the temperature of the ocean surface can increase up to 1–2°C due to the effects of the solar radiation absorption and screening.

In recent decades, microwave radar (SAR) remote sensing imaging techniques have been developed and applied to provide monitoring of oil spills in the ocean (Onstott and Rufenach 1992; Ivanov 2000; Brekke and Solberg 2005; Solberg et al. 2007; Jha et al. 2008; Klemas 2010; Zhang et al. 2011; Salberg et al. 2014; Migliaccio et al. 2015). Although this subject is beyond the scope of this book, we briefly list the main hydrodynamic processes induced by oil spills on water surface. They are the following: (1) advection, (2) turbulent diffusion, (3) surface spreading, (4) vertical mechanical dispersion, (5) emulsification, and (6) evaporation. For more information about oil spill hydrodynamics, we refer the reader to other books (Ehrhardt 2015; Fingas 2015).

2.6 Thermohaline Finestructure

The term "thermohaline finestructure" reflects the most important class of ocean structural heterogeneities associated with vertical profiles of temperature, salinity, and density. The first fundamental investigations of ocean thermohaline structure were made in 1960–1970s and published in the monograph by Fedorov (1978). The inversions of temperature and salinity are presented as high-frequency spatial oscillations (*thermohaline fluctuations*). These experimental data collected in different regions of the world's oceans show that maximal temperature and salinity inversions at a depth of <200–300 m can reach values of 1°C and 0.5 psu. A more detailed and interesting example of thermohaline fluctuations measured in the upper ocean layer (Baltic Sea) is shown in Figure 2.14 (Lips et al. 2008).

Thermohaline fluctuations are produced by natural microturbulence and molecular processes. Usually, the characteristic frequencies of the fluctuations are greater than 1 Hz, and the typical spatial sizes are on the order of several centimeters. The distribution of microfluctuations of temperature, salinity, and density forms a fine deep-ocean stratification or thermohaline finestructure of the ocean. In fact, the thermohaline finestructure exists already and everywhere in different forms in the oceans. The most

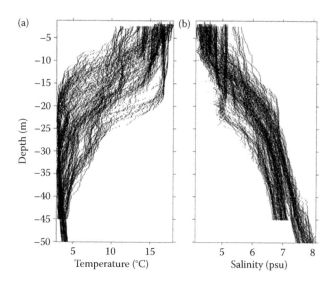

FIGURE 2.14
Fluctuations of thermohaline finestructure in Baltic Sea. Vertical variations of (a) tempera-
ture and (b) salinity. (Adapted from Lips, U. et al. 2008. In *US/EU-Baltic Symposium "Ocean
Observations, Ecosystem-Based Management & Forecasting,"* Tallinn, 27–29 May, 2008. IEEE
Conference Proceedings, pp. 326–333. (Internet search: Bornholm_Taavi_very_final_version.
pptx - BALTEX.)

important role in the transport processes, in particular, heat conductivity
from deep water to the surface, is played by processes of mixing, molecular
diffusion, turbulent diffusion, and convection. As a result of the dynamics of
the heterogeneities, different types of disturbances and instabilities in a deep
ocean may occur. Transportation of these wave disturbances to the ocean
surface and the interactions with wind waves may cause the generation of
two- or three-dimensional hydrodynamic features or anomalies in the fields
of temperature, salinity, roughness, surface current, or even wave breaking.

It was also found that the characteristic time scales or the time of existence
of individual heterogeneities is clearly correlated with corresponding spatial scales. Estimations (Fedorov 1978), made on the basis of known observations, have demonstrated that the average ratio of the scales is on order of
$H/L = 10^{-4} - 10^{-3}$, where H and L are the characteristic vertical and horizontal
dimensions of individual elements of the finestructure. On the other hand,
model estimations show that the relaxation time for the process can change
from tens of hours to tens of days, depending on the scale of the heterogeneities. The presence of dynamic turbulence and microturbulence may also
cause the more rapid formation of thermohaline disturbances, that is, both
the temperature and salinity heterogeneities in the ocean.

Data (Fedorov 1978) show that the inversions of temperature in the finestructure usually appear in local areas. The typical horizontal dimensions lie
within 5–20 km with the thickness of individual heterogeneities of 5–20 m.

A horizontal temperature pattern ("microsurvey") is observed in the main thermocline (a depth range of 140–170 m). Microsurvey presents a set of temperature contours that sometimes have the character of deep-water thermohaline fronts. The magnitude of temperature increases with depth within the main contour (Fedorov 1978). Similar configurations were registered at different levels of ocean depth in areas of the main thermocline. Zones of high temperature and salinity gradients on an isopycnal surface ($\sigma_t = 25.00$) represent the narrow thermohaline fronts that can change with time.

Statistical characteristics of thermohaline finestructure have been investigated as well (Fedorov 1978). The experimental spectral density of temperature and salinity fluctuations have a power law and are approximately proportional to $\Psi_{t-s} \sim K^{-2}$. Such spectra correspond to the existence of numerous sharp deviations of the temperature and salinity stratification. It is important to note that the low-frequency part of the thermohaline spectra with the scale >25–30 m has features associated with the nonstationary and kinematic effect of internal waves. The high-frequency part of the spectra reflects a stable thermohaline structure, especially temperature inversions. Vertical turbulent mixing is a general mechanism of formation of the vertical quasi-uniform layers with a thickness of 5–10 m and less. Their horizontal scales can change from a few hundreds of meters to 1–5 km.

In general, the temporal dynamics of the thermohaline heterogeneities, fields of temperature T and salinity S, are described by equations of heat and salt balance in incompressible liquid:

$$\frac{\partial T}{\partial t} + U\nabla T = k_T \nabla^2 T, \tag{2.33}$$

$$\frac{\partial S}{\partial t} + U\nabla S = k_S \nabla^2 S, \tag{2.34}$$

where U(x,y,z) is the field of velocity, k_T is the coefficient of molecular conductivity of heat, and k_S is the coefficient of salt diffusion.

Equations 2.33 and 2.34 reflect the processes of formation and evolution of thermohaline structure during all types of motion (stationary, nonstationary, turbulent, molecular) which achieve local balance of the vertical and horizontal fluxes of heat and salt. These equations must be considered along with the Navier–Stokes equations:

$$\frac{\partial U}{\partial t} + (U \cdot \nabla)U = -\frac{1}{\rho_t}\nabla P + V_0 \nabla^2 U + F, \tag{2.35}$$

$$\nabla \cdot U = 0,$$

$$\frac{\partial \rho_t}{\partial t} = 0,$$

where U is the velocity, t is the time, P is the pressure, ρ_t is the density, v_0 is the viscosity, and F is the forces term (gravity, stirring).

The system of Equations 2.33 through 2.35 is rewritten in terms of fluctuating components. For example, the following form is used: $U = <U> + U'$, $T = <T> + T'$, $S = <S> + S'$, where brackets $<>$ denote averaging at an ensemble, and prime denotes the fluctuating part of the parameter. The system of dynamical equations can be used for the numerical modeling of free thermohaline double-diffusive convection and evolution of two- or three-dimensional thermohaline patterns in the oceans.

2.7 Double-Diffusive Convection and Instabilities

Double-diffusive convection, or double diffusion, is a kind of convection in the ocean, which originates from the difference in molecular diffusivities for heat and salt. Depending on mean temperature and salinity stratification, there are two types of double-diffusive convection in the ocean: salt finger convection and diffusive convection (Turner 1974; Schmitt 1994; Brandt and Fernando 1995; Radko 2013).

The effect of destabilization of the original stable stratification and motion is caused by an inequality in the rates of molecular diffusion of momentum and mass. This phenomenon is well known in oceanography. An example is the development of circular baroclinic vortex in deep ocean with specific stratification of the density field. The diffusion phase begins during the deformation of the density field by turbulent motions. Since the mechanism of turbulence already does not depend on the molecular diffusion rates of heat and salt, the process concentrates in a thin transitional sheet (pycnocline), separating two uniform layers with constant values of density and velocity.

It is assumed that the mechanism of forming thermohaline finestructure is associated with the redistribution of the potential energy between the saline and thermal components of meso- and macro-scale stratification. The process of double diffusion is characterized by the coefficient of molecular conductivity of heat k_T (the average value for oceanic conditions is $1.4 \cdot 10^{-3}$ cm^2/s) and by the coefficient of molecular diffusion of salt k_S (a typical value is $1.3 \cdot 10^{-5}$ cm^2/s). The convection is the main physical mechanism of energy exchange between saline and thermal stratification.

In the simplest hydrostatic approach for ocean–water state, the relationship between deviation of density $\Delta\rho = \rho - \rho_0$ and the deviations of temperature $\Delta T = T - T_0$ and salinity $\Delta S = S - S_0$ in thermohaline structure is described by the following linear equation (Fedorov 1978):

$$\Delta\rho = -\alpha\Delta T + \beta\Delta S, \qquad (2.36)$$

where ρ_0, T_0, and S_0 are the initial values of density, temperature, and salinity, respectively;

$$\alpha = -\frac{1}{\rho_0}\left(\frac{\partial \rho}{\partial T}\right)_{S,P}$$

is the gradient evaluated at fixed values of salinity and pressure (S, P), and

$$\beta = \frac{1}{\rho_0}\left(\frac{\partial \rho}{\partial S}\right)_{T,P}$$

is the gradient evaluated at fixed values of temperature and pressure (S, P).

In the case where the original fields of temperature T(x,y,z) and salinity S(x,y,z) are compensated by each other, density heterogeneities do not arise. In the other case, density heterogeneities occur, and variations of density and pressure must contribute to the development of local motions U(x,y,z). Thus, the process can produce nonstationary motions in a stratified ocean, including the development of convective instability due to the influence of horizontal heterogeneities of temperature and salinity.

The criteria of thermohaline (in)stability as a background of the development of double-diffusive convection are estimated by the dimensionless density ratio. Two forms of the density ratio are used (Fedorov 1978):

$$R_\rho = \frac{\alpha\Delta T}{\beta\Delta S} \tag{2.37}$$

or

$$R_\rho = \frac{\beta\Delta S}{\alpha\Delta T}. \tag{2.38}$$

This ratio is based on Equation 2.36 and written in such a way that the stable thermohaline properties are described by the numerator, and the unstable properties are described by the denominator.

In the case of neutral stratification, obviously, $R_\rho = 1$. If the vertical distribution of temperature is stable $\Delta T/\Delta z < 0$, and the unstable contribution to the density stratification is introduced by the vertical distribution of salinity $\Delta S/\Delta z < 0$, the ratio must be used in the form of Equation 2.37. In the case of temperature inversion $\Delta T/\Delta z > 0$, partially or completely stabilized by salinity distribution $\Delta S/\Delta z > 0$, the ratio must be used in the form of Equation 2.38.

Figure 2.15 illustrates different types of thermohaline convection (Zhurbas. *Lecture on Oceanography*. https://www.yumpu.com/en/document/

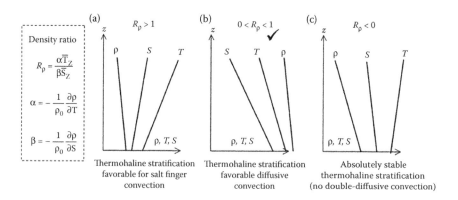

FIGURE 2.15
Types of thermohaline convection. (a) Thermohaline stratification favorable for (a) salt finger convection and (b) diffusive convection. (c) Absolutely stable thermohaline stratification (no double-diffusive convection). (Adapted from Zhurbas. Lecture on Oceanography. Internet search http://msi.ttu.ee/~elken/Zhurbas_L08.pdf.)

view/33829134/lecture-6-oceanic-fine-structure/7). Laboratory experiments with thermohaline convection (Turner 1973, 1978) and their numerical approximation (Fedorov 1978) show that the ratio of fluxes of heat and salt k_S/k_T during layered convection in the presence of two-layer stratification is changed exponentially with degreasing of values of $R_\rho = \beta\Delta S/\alpha\Delta T$ from 1 to 7.

Two regimes of thermohaline convection are observed. The first regime is "constant" when the ratio of mass fluxes is equal to 0.15 and density ratio $R_\rho = \beta\Delta S/\alpha\Delta T > 2$. The second regime is "variable" with the ratio of mass fluxes ranges from 1.85 to 0.85 and density ratio $1 < R_\rho = \beta\Delta S/\alpha\Delta T < 2$. Constant regime of layered convection is formed due to the influence of temperature and salinity gradients that are adjusted for equilibrium. There is some opinion that the existence of the constant regime is the typical case also for the development of the salt fingers in the ocean.

Salt fingers are a form of cellular convection, developed in a two-component liquid medium with a stable density stratification (Fedorov 1978; Charru 2011). They are produced by the combination of the stabilizing contribution of the vertical temperature gradient and the destabilizing contribution of the vertical salinity gradient. The salt fingers are convective cells, elongated vertically. According to experimental data (Schmitt 2003; Huang 2009), they have a square cross section with sides up to 0.4 cm and are several centimeters in length. The salt fingers are associated with a significant downward flux of heat. The generation of vertically periodic convective cells, leading to the formation of salt fingers, can occur due to local gradients of temperature in a stable salinity-stratified fluid. Laboratory experiments (Popov and Chashechkin 1979; Taylor and Buchens 1989; Taylor 1993; Taylor and Veronis 1996; Schmitt 2003) show that the vertical dimension of the convective cells increases linearly with the overheating temperature. The size of the observed cells varied from 0.4 to 1.2 cm in the range of the overheating temperature 0–4°C.

During the development of salt fingers in an ocean layer with strong vertical gradients of temperature and salinity, the viscous forces may have the same order of magnitude as the buoyancy forces produced by double-diffusive convection. It is known that the general criteria for dynamical instability, that is, for the appearance of turbulence in the viscosity flow of liquid, is the Reynolds number $Re = U_0(L/v)$, where U_0 is the characteristic velocity of flow, L is the characteristic linear scale of the motion, and v is the kinematic viscosity. For example, for laminar flow, $Re \sim 2000$. A certain value of Re, corresponding to the point of dynamical instabilities, is named the critical value of the Reynolds number Re_{crit}. The regime is laminar if $Re > Re_{crit}$ and the regime is turbulent if $Re > Re_{crit}$. However, the Reynolds number is not a sufficient criterion for the original turbulence in the ocean. The important role in the occurrence of ocean turbulence is played by vertical profiles of temperature and salinity as well as density stratification. Therefore, other useful criterion is defined by the Richardson number $Ri = (g/\rho_0)(\partial\rho/\partial z)(\partial U/\partial z)^{-2}$. In stationary plane-parallel flow with shear, the value of the critical Richardson number, when turbulent regime occurs, equals approximately $Ri_{crit} = 1/4$. The regime with $Ri < Ri_{crit} \approx 1/4$ corresponds to the appearance of hydrodynamical instability (e.g., development of vortex disturbances begins at values $Ri = 0.05$–0.1). Field experiments have shown that the convective instability can appear under conditions $Ri < Ri_{crit} \approx 1$–2 (Fedorov 1978).

In the ocean, thermohaline intrusions can cause the formation of large-scale stepped finestructure of the main thermocline. For example, at a depth of 200–500 m, the "individual" steps with a thickness from 8 to 55 m and horizontal size of at least 35 km were measured (Zhurbas and Ozmidov 1983, 1984). The density ratio in the stepped structure does not depend on the depth (averaging $R_\rho = 1.62$). Such stepped erosion of the main thermocline in the ocean has a convective nature related to double-diffusion effects.

At the present time, the theory of formation of salt fingers and deep water stepped structure in the ocean under the influence of thermohaline processes is not well-enough advanced. Obviously, dynamical models for the numerical analysis of double-diffusive convection must be three-dimensional, unlike the case of free convection. For example, the effect of the development of free convection in stratified cooling seawater is well explained by a two-dimensional mathematical model (Bune et al. 1985).

2.8 Self-Similarity and Turbulent Intrusions

The analysis of a large body of field measurements shows the universal nature of the temperature and salinity distributions in the thermocline. In this context, a new concept about the structure of temperature, salinity, and density fields in the ocean's upper thermocline was suggested on the basis of the self-similarity hypothesis (Barenblatt 1978b, 1996). In this theory, a possibility of the existence of a stationary temperature or density wave is considered. An analytic solution of the linear transport equation for the excess temperature was found in the approximation of small-scale, homogeneous, and stationary motions that determine the transport mechanism. As a result, the vertical exchange coefficient in the upper thermocline is estimated.

The calculated exchange coefficient has a value larger than the molecular coefficient and at the same time smaller than the coefficient of turbulent diffusion. This means that a turbulent erosion of the thermocline may exist, which is associated with effects of double-diffusive convection or breaking internal waves. Later, the two-phase model of unsteady turbulent heat and mass transfer in the upper oceanic thermocline was developed (Barenblatt 1982). The liquid was represented as a hydro-mechanical ensemble of two penetrating phases: turbulent spots and laminar sheets separating them. The model explains the existence of temperature jumps in the thermocline and the appearance of stepped structure. In principle, the model may be used to describe the heat–salt mass balance in the presence of salt fingers and temperature inversions.

Temperature jumps or fronts in the ocean's upper layer are a possible reason for the development of oscillatory motions and generation of internal waves. In these cases, an effect of the amplification of internal wave amplitude can occur, which may cause the development of double-diffusive convection. The theory of wave fronts in dissipating media is well known: these are shock waves in gases, collisionless fronts in plasma, and electromagnetic shock waves in solid. However, dispersion may significantly influence the behavior of wave fronts, in particular, oceanic fronts. A model of possible wave structure in the stratified ocean was developed recently on the basis of the Korteweg-de Vries–Burgers equation (Barenblatt and Shapiro 1984). It was found that the solution has a "traveling wave front" form at a certain choice of the relation between dispersion and viscosity coefficients. Oscillations arise behind the front. Such front was named "dispersion front," unlike shock waves and smoothed steps. In fact, the theory has predicted the appearance of turbulent flow-induced *preordering* or precursor.

Other specific turbulent features—turbulent spots or intrusions—are associated with strong density stratification in the ocean (Monin and Ozmidov 1985; Baumert et al. 2005). There are a number of sources of the generation of turbulent spots within the near-surface transition boundary layers; among them, the most important are (1) wind wave breaking is accompanied by the formation of intermittent turbulent spots; (2) breaking of internal waves in a shear flow; (3) hydro-acoustic action associated with pulsed disturbances induced by sound waves, mechanical or electrical means; and (4) collapse of deep ocean turbulent wake.

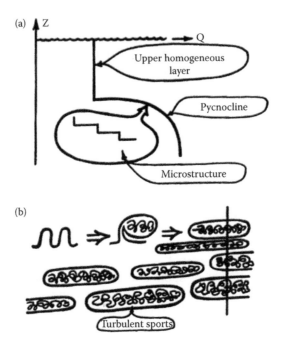

FIGURE 2.16

Model illustration of turbulence under the strong density stratification. (a) Oceanic microstructure in the upper pycnocline and (b) turbulent spot. (Original drawing from Barenblatt, G. I. 1992. In *Proceedings of Second International Conference on Industrial and Applied Mathematics*. R. E. O'Mailey (ed.). Society for Industrial & Applied Mathematics SIAM, pp. 15–29.)

The following description of what turbulent spots look like was done by Barenblatt (1978a): "turbulence…has an unusual spatial structure; it is concentrated in pancake layers which extend in the horizontal direction over a distance that significantly exceeds their thickness." According to the hydromechanic model (Barenblatt 1978a, 1991, 1992), there is a peculiar microstructure in the upper pycnocline (i.e., in the layer with sharp density gradient) due to spatial intermittency of turbulence. Turbulence under the strong density stratification is concentrated in turbulent spots, squeezed by ambient nonturbulent fluid. Figure 2.16 taken from paper (Barenblatt 1992) explains the ocean microstructure and the formation of turbulent spots in the upper pycnocline. Such spots occur due to breaking of internal waves in a shear flow of stably stratified ocean.

Dynamics of turbulent spots is accompanied by the changes of the spot thickness h that can be described by the equation (Barenblatt 1978a).

$$\frac{\partial h}{\partial t} = k \Delta h^5, \tag{2.39}$$

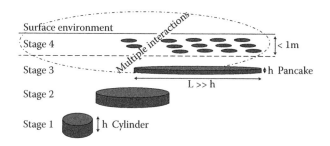

FIGURE 2.17
Structure and dynamics of turbulent spots in the ocean. Schematic diagram illustrates the following four stages, the first three of which are based on the Barenblatt's model: initial stage #1; intermediate stationary stage #2; final viscous stage #3; and interaction stage #4. h is the thickness and L is the length of a single spot.

where t is the time, Δ is the horizontal Laplace operator, and k is a constant that depends on ocean stratification. The thickness of the spot is decreases slowly and its radius increases very slowly:

$$h_0(t) \sim (t-t_0)^{-1/5}, \quad r_0(t) \sim (t-t_0)^{1/10}, \tag{2.40}$$

where h_0 is the maximal spot thickness and r_0 is the spot radius. Thus, the turbulent spot on the intermediate asymptotic stage has a form of a thin disk.

Figure 2.17 illustrates the transformation of turbulent spots in the upper ocean layer at different stages. Naturally, three following stages are considered (Barenblatt 1978a): (#1) initial stage of the free intrusion; (#2) intermediate stationary stage; and (#3) final viscous stage. We add the fourth, interaction stage (#4), which describes collapse (or multiple split) of large-scale thin turbulent spot (L~10–30 m length) into a number of mid- or small-scale turbulent spots (L~10–20 cm length) due to interactions with the surface environment (wind, surface waves, currents, etc.). Stage (#4) may cause the generation of vortex-like coherent structure or other turbulent features at the air–sea interface, which are potentially detectable using high-resolution passive microwave imagery.

Turbulent spots play an important role in the interactions between submerged turbulence and surface waves. Microstructure-turbulence spots create a homogeneous layer with conditions favorable for the development of modulational instabilities, vorticity anomalies, or small-scale eddies that may affect the surface wave spectrum. As a result, transformations or excitations of certain spectral components or their groups may occur that can also be detected by multiband passive microwave radiometer.

In this context, at least the three following thermohaline mechanisms effecting the change of microwave signal from the ocean surface can be considered:

1. Generation of small-scale periodic gratings and/or coherent cells in the roughness field

2. Appearance of turbulent spots and roughness anomalies with specific distribution of hydro-physical characteristics

3. Change of integral heat and mass fluxes through the disturbed air–water interface

The first case corresponds to the appearance of a set of high-frequency harmonics in the ambient wave number spectrum due to the periodic redistribution of the small-scale roughness components. Selective multimode excitations of the spectrum are accompanied by angular redistribution of spectral energy density. The effect has a local character and may exist for a short time.

The second case causes the deformation of the spectrum as a whole. For example, it may be a change of power index (exponent). Deformations of the capillary part of wave number spectrum can arise from turbulent spots due to influence of higher horizontal gradient of temperature (salinity) causing a change in surface tension coefficient. However, capillary effects are very unstable and slow.

Finally, the third case reflects the change of the *thermo-hydrodynamical* condition of the ocean–atmosphere boundary layer. For example, the occurrence of energetically active zones in the ocean is very probably in the cases of the development of "crisis" atmosphere situations such as typhoon, a tropical cyclone vortex, and/or cyclonic eddy. Their surface manifestations have been observed by microwave radiometer in field experiments (Cherny and Raizer 1998).

2.9 Summary

In this chapter, we made a survey of ocean processes and phenomena, which are of great interest and importance in remote sensing. Hydro-physical factors considered above are the main components of the ocean–atmosphere system; they provide a broadband electromagnetic response at microwave frequencies.

Some of environmental factors—wind waves, surface roughness, breaking waves, foam, whitecap, bubbles, spray, and dense aerosol—contribute directly to ocean emissivity, causing measurable variations of microwave emission. Other factors—turbulent spots, intrusions, thermohaline and double-diffusive processes, as well as subsurface bubbly flows, jet streams, and wakes—can be considered as possible indicators of deep ocean processes. Under certain conditions, these factors trigger the development of surface disturbances and/or multiple interactions. Therefore, they may contribute indirectly to ocean emissivity through the changes in dynamic, structural, and statistical characteristics of the air–sea interface.

Concerning applied hydrodynamics related to remote sensing, a number of important scientific problems remain unsolved. Among them such natural phenomena as wave instabilities, bifurcations, and nonlinear wave interactions in the ocean further deserve more detailed consideration. The generation and evolution of two-dimensional coherent (self-similar) hydrodynamic structures and complex patterns in the ocean are also poorly studied phenomena. These problems require our attention as well.

Transport processes and turbulence in the stratified ocean and the corresponding surface manifestations of deep ocean events are the most important issues in context with recent remote sensing observations. These and other critical for microwave studies processes in the upper ocean need to be further investigated in laboratory and field experiments as well as using theoretical and numerical methods of applied hydrodynamics.

References

Abe, T. 1963. In situ formation of stable foam in sea water to cause salty wind damage. *Papers in Meteorology and Geophysics*, 14(2):93–108.

Albert, M. F. M. A., Anguelova, M. D., Manders, A. M. M., Schaap, M., and de Leeuw, G. 2015. Parameterization of oceanic whitecap fraction based on satellite observations. *Atmospheric Chemistry and Physics Discussion*, 15:21219–21269. Doi: 10.5194/acpd-15-21219-2015.

Alpers, W., Blume, H.-J. C., Garrett, W. D., and Hühnerfuss, H. 1982. The effect of monomolecular surface films on the microwave brightness temperature of the sea surface. *International Journal of Remote Sensing*, 3(4):457–474. Doi: 10.1080/01431168208948415.

Alpers, W. and Hühnerfuss, H. 1989. The damping of ocean waves by surface films: A new look at an old problem. *Journal of Geophysical Research*, 94(C5):6251–6265.

Andreas, E. L. 1992. Sea spray and the turbulent air-sea heat fluxes. *Journal of Geophysical Research*, 97(C7):11429–11441.

Andreas, E. L., Edson, J. B., Monahan, E. C., Rouault, M. P., and Smith, S. D. 1995. The spray contribution to net evaporation from the sea: A review of recent progress. *Boundary-Layer Meteorology*, 72:3–52.

Andreas, E. L., Jones, K. F., and Fairall, C. W. 2010. Production velocity of sea spray droplets. *Journal of Geophysical Research*, 115:C12065.

Anguelova, M., Barber Jr. R. P., and Wu, J. 1999. Spume drops produced by the wind tearing of wave crests. *Journal of Physical Oceanography*, 29:1156–1165.

Anguelova, M. D., Gaiser, P. W., and Raizer, V. 2009. Foam emissivity models for microwave observations of oceans from space. In *Proceedings of International Geoscience and Remote Sensing Symposium*, July 12–17, 2009, Cape Town, South Africa, Vol. 2, pp.: II-274–II-277. Doi: 10.1109/IGARSS.2009.5418061.

Anguelova, M. D. and Huq, P. 2012. Characteristics of bubble clouds at various wind speeds. *Journal of Geophysical Research*, 117:C03036.

Anguelova, M. D. and Webster, F. 2006. Whitecap coverage from satellite measurements: A first step toward modeling the variability of oceanic whitecaps. *Journal of Geophysical Research*, 111:C03017.

Apel, J. R. 1987. *Principles of Ocean Physics (International Geophysics Series, Vol. 38)*. Academic Press, London, UK.

Apel, J. R. 1994. An improved model of the ocean surface wave vector spectrum and its effects on radar backscatter. *Journal of Geophysical Research*, 99(C8):16269–16291.

Babanin, A. 2011. *Breaking and Dissipation of Ocean Surface Waves*. Cambridge University Press, Cambridge, UK.

Badulin, S. I., Pushkarev, A. N., Resio, D., and Zakharov, V. E. 2005. Self-similarity of wind-driven seas. *Nonlinear Processes in Geophysics*, 12:891–945.

Bakhanov, V. V. and Ostrovsky, L. A. 2002. Action of strong internal solitary waves on surface waves. *Journal of Geophysical Research*, 107(C10):3139.

Baldy, S. 1988. Bubbles in the close vicinity of breaking waves: Statistical characteristics of the generation and dispersion mechanism. *Journal of Geophysical Research*, 93(C7):8239–8248.

Banner, M. L. and Morison, R. P. 2010. Refined source terms in wind wave models with explicit wave breaking prediction. Part I: Model framework and validation against field data. *Ocean Modelling*, 33:177–189. Doi: 10.1016/j.ocemod.2010.01.002.

Barenblatt, G. I. 1978a. Dynamics of turbulent spots and intrusions in a stably stratified fluid. *Izvestiya Atmosphere and Oceanic Physics*, 14(2):139–145 (translated from Russian).

Barenblatt, G. I. 1978b. Self-similarity of temperature and salinity distributions in the upper thermocline. *Izvestiya Atmosphere and Oceanic Physics*, 14(11):820–823 (translated from Russian).

Barenblatt, G. I. 1982. A model of non steady turbulent heat and mass transfer in a liquid with highly stable stratification. *Izvestiya Atmosphere and Oceanic Physics*, 18(3):201–205 (translated from Russian).

Barenblatt, G. I. 1991. Dynamics of turbulent spots in stably stratified fluid. In *Mathematical Approaches in Hydrodynamics*. T. Miloh (ed.). Society for Industrial & Applied Mathematics, pp. 373–381.

Barenblatt, G. I. 1992. Intermediate asymptotics in micromechanics. In *Proceedings of Second International Conference on Industrial and Applied Mathematics*. R. E. O'Mailey (ed.). Society for Industrial & Applied Mathematics SIAM, July 8–12, 1991, Washington, DC, USA, pp. 15–29.

Barenblatt, G. I. 1996. *Scaling, Self-Similarity, and Intermediate Asymptotics*. Cambridge University Press, Cambridge, UK.

Barenblatt, G. I., Leykin, I. A., Kaz'min, A. S., Kozlov, V. A., Raszhivin, V. A., Fillippov, I. A., Frolov, I. D., and Chuvil'chikov, S. I. 1985. Sooloy (Suloy) in the White Sea. *Doklady, USSR Academy of Sciences (Doklady Akademii Nauk SSSR)*, 281(6):1435–1439 (in Russian).

Barenblatt, G. I., Leykin, I. A., Kaz'min, A. S., Kozlov, V. A., Raszhivin, V. A., Fillippov, I. A., Frolov, I. D., and Chuvilchikov, S. I. 1986a. Sooloy (suloy) in White Sea. Part 1. Observations of the sooloy, and its connection with the tidal currents. *Morskoy Gidrofizichesky Zhurnal (Marine Hydrophysical Journal)*, (2):49–53 (in Russian).

Barenblatt, G. I., Leykin, I. A., Kaz'min, A. S., Kozlov, V. A., Raszhivin, V. A., Fillippov, I. A., Frolov, I. D., and Chuvilchikov, S. I. 1986b. Sooloy (suloy) in White Sea. Part 2. The Sooloy's connection with the local bottom relief, and the wave measuring. *Morskoy Gidrofizichesky Zhurnal (Marine Hydrophysical Journal)*, (5):29–33 (in Russian).

Barenblatt, G. I. and Shapiro, G. I. 1984. A contribution to the theory of wave front structure in dispersive dissipating media. *Izvestiya Atmosphere and Oceanic Physics*, 20(3):210–215 (translated from Russian).

Basovich, A. Y. 1979. Transformation of the surface wave spectrum due to the action of an internal wave. *Izvestiya, Atmospheric and Oceanic Physics*, 15(6):448–452 (translated from Russian).

Basovich, A. Ya. and Talanov, V. I. 1977. Transformation of the spectrum of short surface waves on inhomogeneous currents. *Izvestiya, Atmospheric and Oceanic Physics*, 13(7):514–519 (translated from Russian).

Baumert, H. Z., Simpson, J. H., and Sündermann, J. 2005. *Marine Turbulence: Theories, Observations, and Models*. Cambridge University Press, Cambridge, UK.

Beal, R. C., Tilley, D. C., and Monaldo, F. M. 1983. Large- and small-scale evolution of digitally processed ocean wave spectra from SEASAT synthetic aperture radar. *Journal of Geophysical Research*, 88(C3):1761–1778.

Benilov, A. Yu. 1973. The turbulence generation in the ocean by surface waves. *Izvestiya, Atmospheric and Oceanic Physics*, 9(3):160–164 (translated from Russian).

Benilov, A. Yu. 1991. *Soviet Research of Ocean Turbulence and Submarine Detection*. Delphic Associates Inc.

Bikerman, J. J. 1973. *Foams*. Springer, Berlin.

Blanchard, D. C. 1963. The electrification of the atmosphere by particles from bubbles in the sea. In *Progress in Oceanography*. M. Sears (ed.), pp. 73–202. Pergamon Press, New York. Doi: 10.1016/0079-6611(63)90004-1.

Blanchard, D. C. 1971. Whitecap at sea. *Journal of the Atmospheric Sciences*, 28(4): 645–651.

Blanchard, D. C. 1983. The production, distribution, and bacterial enrichment of the sea-salt aerosol. In *The Air–Sea Exchange of Gases and Particles*. P. S. Liss and W. G. M. Slinn (eds.), pp. 407–454. Kluwer, D. Reidel, Dordrecht, The Netherlands.

Blanchard, D. C. 1990. Surface-active monolayers, bubbles, and jet drops. *Tellus B*, 42:200–205.

Bobak, J. P., Asher, W. E., Dowgiallo, D. J., and Anguelova, M. D. 2011. Aerial radiometric and video measurements of whitecap coverage. *IEEE Transactions on Geoscience and Remote Sensing*, 49(6):2183–2193.

Bondur, V. G. and Sharkov, E. A. 1982. Statistical properties of whitecaps on a rough sea. *Oceanology*, 22(3):274–279 (translated from Russian).

Bortkovskii, R. S. 1987. *Air–Sea Exchange of Heat and Moisture during Storms*. D. Reidel, Dordrecht, The Netherlands.

Bortkovskii, R. S. and Novak, V. A. 1993. Statistical dependencies of sea state characteristics on water temperature and wind-wave age. *The Journal of Marine Systems*, 4(2):161–169.

Bowyer, P. A. 2001. Video measurements of near-surface bubble spectra. *Journal of Geophysical Research*, 106(C7):14179–14190.

Brandt, A. and Fernando, H. J. S. 1995. *Double-Diffusive Convection (Geophysical Monograph Series, Vol. 94)*. American Geophysical Union, Washington DC.

Brekke, C. and Solberg, H. A. 2005. Oil spill detection by satellite remote sensing. *Remote Sensing of Environment*, 95(1):1–13.

Breward, C. J. 1999. The mathematics of foam. Ph.D. dissertation. University of Oxford. https://core.ac.uk/download/pdf/96508.pdf

Bune, A. V., Ginzburg, A. I., Polezhaev, V. I., and Fedorov, K. N. 1985. Numerical and laboratory modeling of the development of convection in a water layer cooled from the surface. *Izvestiya Atmosphere and Oceanic Physics*, 21(9):736.

Bunner, M. L. and Peregrine, D. H. 1993. Wave breaking in deep water. *Annual Review of Fluid Mechanics*, 25:373–397.

Callaghan, A., de Leeuw, G., Cohen, L., and O'Dowd, C. D. 2008. Relationship of oceanic whitecap coverage to wind speed and wind history. *Geophysical Research Letters*, 35(23):L23609.

Callaghan, A. H., Deane, G. B., and Stokes, M. D. 2013. Two regimes of laboratory whitecap foam decay: Bubble plume controlled and surfactant stabilized. *Journal of Physical Oceanography*, 43:1114–1126.

Callaghan, A. H., Deane, G. B., Stokes, M. D., and Ward, B. 2012. Observed variation in the decay time of oceanic whitecap foam. *Journal of Geophysical Research*, 117:C09015.

Callaghan, A. H. and White, M. 2009. Automated processing of sea surface images for the determination of whitecap coverage. *Journal of Atmospheric and Oceanic Technology*, 26(2):383–394.

Chalikov, D. and Babanin, A. V. 2012. Simulation of wave breaking in one-dimensional spectral environment. *Journal of Physical Oceanography*, 42(11):1745–1761. Doi: 10.1175/JPO-D-11-0128.1.

Charru, F. 2011. *Hydrodynamic Instabilities (Cambridge Texts in Applied Mathematics)*. University Press, Cambridge, UK.

Chen, G., Kharif, C., Zaleski, S., and Li, J. 1999. Two-dimensional Navier–Stokes simulation of breaking waves. *Physics of Fluids*, 11(1):121–133.

Chen, Y., Guza, R. T., and Elgar, S. 1997. Modeling spectra of breaking surface waves in shallow water. *Journal of Geophysical Research*, 102(C11):25035–25046.

Cherny, I. V. and Raizer, V. Y. 1998. *Passive Microwave Remote Sensing of Oceans*. Wiley, Chichester, UK.

Clift, R., Grace, J. R., and Weber, M. E. 1978. *Bubbles, Drops, and Particles*. Academic Press, New York.

Cox, C. and Munk, W. 1954. Statistics of the sea surface derived from sun glitter. *Journal of Marine Research*, 13(2):199–227.

Craik, A. D. D. 1985. *Wave Interactions and Fluid Flows*. Cambridge University Press, Cambridge.

Creamer, D. B. and Wright, J. A. 1992. Surface films and wind wave growth. *Journal of Geophysical Research*, 97(C4):5221–5229.

da Silva, J. C. B., Ermakov, S. A., Robinson, I. S., Jeans, D. R. G., and Kijashko, S. V. 1998. Role of surface films in ERS SAR signatures of internal waves on the shelf: 1. Short-period internal waves. *Journal of Geophysical Research*, 103(C4):8009–8031.

Donelan, M. A. and Pierson Jr., W. J. 1987. Radar scattering and equilibrium range in wind-generated waves. *Journal of Geophysical Research*, 92(C5):4971–5029.

Drazin, P. G. 2002. *Introduction to Hydrodynamic Stability (Cambridge Texts in Applied Mathematics)*. Cambridge University Press, Cambridge.

Ehrhardt, M. 2015. *Mathematical Modelling and Numerical Simulation of Oil Pollution Problems*. Springer, Switzerland.

Elfouhaily, T., Chapron, B., Katsaros, K., and Vandemark, D. 1997. A unified directional spectrum for long and short wind-driven waves. *Journal of Geophysical Research*, 102(15):781–796.

Engelbrecht, J. 1997. *Nonlinear Wave Dynamics: Complexity and Simplicity*. Kluwer Academic Publishers, Dordrecht, Boston, London.

Ermakov, S. A. 2010. On the intensification of decimeter-range wind waves in film slicks. *Izvestiya, Atmospheric an Oceanic Physics*, 46(2):208–213 (translated from Russian).

Ermakov, S. A., Salashin, S. G., and Panchenko, A. R. 1992. Film slicks on the sea surface and some mechanisms of their formation. *Dynamics of Atmospheres and Oceans*, 16(3–4):279–304.

Espedal, H. A., Johannessen, O. M., and Knulst, J. 1996. Satellite detection of natural film on the ocean surface. *Geophysical Research Letters*, 23(22):3151–3154.

Etkin, V., Raizer, V., Stulov, A., and Zhuravlev, K. 1995. Airborne optical measurements of wind-wave spectral perturbations induced by ocean internal waves. In *Proceedings of Combined Optical-Microwave Earth and Atmosphere Sensing*, April 3–6, 1995, Atlanta, pp. 81–83. Doi: 10.1109/COMEAS.1995.472333.

Faber, T. E. 1995. *Fluid Dynamics for Physicists*. Cambridge University Press, Cambridge, UK.

Fairall, C. W., Kepert, J. D., and Holland, G. J. 1994. The effect of sea spray on surface energy transports over the ocean. *Global Atmosphere and Ocean System*, 2(2–3):121–142.

Farmer, D. M. and Lemone, D. D. 1984. The influence of bubbles on ambient noise in the ocean at high wind speed. *Journal of Physical Oceanography*, 14(11): 1762–1778.

Fedorov, K. N. 1978. *The Thermohaline Finestructure of the Ocean*. Pergamon Press, Oxford.

Fedorov, K. N. and Ginsburg, A. I. 1992. *The Near-Surface Layer of the Ocean*. VSP, Utrecht, The Netherlands.

Fingas, M. 2015. *Handbook of Oil Spill Science and Technology*. Wiley, Hoboken, NJ.

Gade, M., Byfield, V., Ermakov, S., Lavrova, O., and Mitnik, L. 2013. Slicks as indicators for marine processes. *Oceanography*, 26(2):138–149.

Gade, M., Hühnerfuss, H., and Korenowski, G. M. 2006. *Marine Surface Films: Chemical Characteristics, Influence on Air-Sea Interactions and Remote Sensing*. Springer, Berlin, Heidelberg.

Garrett, W. D. 1967. Stabilization of air bubbles at the air-sea interface by surface-active material. *Deep Sea Research*, 14:661–672.

Garrett, W. D. 1968. The influence of monomolecular surface films on the production of condensation nuclei from bubbled sea water. *Journal of Geophysical Research*, 73(16):5145–5150.

Gasparovic, R. F., Apel, J. R., and Kasischke, E. S. 1988. An overview of the SAR internal wave signature experiment. *Journal of Geophysical Research*, 93(C10):12304–12316.

Gemmrich, J. R., Zappa, C., Banner, M. L., and Morison, R. P. 2013. Wave breaking in developing and mature seas. *Journal of Geophysical Research*, 118(9):4542–4552.

Glasman, R. E. 1991a. Statistical problems of wind-generated gravity waves arising in microwave remote sensing of surface winds. *IEEE Transactions on Geoscience and Remote Sensing*, 29(1):135–142.

Glasman, R. E. 1991b. Fractal nature of surface geometry in a developed sea. In *Non-Linear Variability in Geophysics*. D. Schertzer and S. Lovejoy (eds.). Kluwer Academic Publishers, Dordrecht, pp. 217–226.

Grue, J., Gjevik, B., and Weber, J. E. 1996. *Waves and Nonlinear Processes in Hydrodynamics*. Kluwer Academic Publisher, Dordrecht, The Netherlands.

Grushin, V. A., Raizer, V. Y., Smirnov, A. V., and Etkin, V. S. 1986. Observation of nonlinear interaction of gravity waves by optical and radar techniques. *Doklady of Russian Academy of Sciences*, 290(2):458–462 (in Russian).

Grythe, H., Ström, J., Krejci, R., Quinn, P., and Stohl, A. 2014. A review of sea-spray aerosol source functions using a large global set of sea salt aerosol concentration measurements. *Atmospheric Chemistry and Physics*, 14:1277–1297.

Guan, C., Hu, W., Sun, J., and Li, R. 2007. The whitecap coverage model from breaking dissipation parametrizations of wind waves. *Journal of Geophysical Research*, 112:C05031.

Hanson, J. L. and Phillips, O. M. 1999. Wind sea growth and dissipation in the open ocean. *Journal of Physical Oceanography*, 29(8):1633–1648.

Hasselman, K. 1962. On the nonlinear energy transfer in a gravity wave spectrum. *Journal of Fluid Mechanics*, 12:481–500.

Hasselman, K. 1974. On spectral dissipation of ocean waves due to whitecapping. *Boundary-Layer Meteorology*, 6:107–127

Herbert, T. 1988. Secondary instability of boundary layers. *Annual Review of Fluid Mechanics*, 20:487–526.

Higuera, P., del Jesus, M., Lara, J. L., Losada, I. J., Guanche, Y., and Barajas, G. 2013. Numerical simulation of three-dimensional breaking waves on a gravel slope using a two-phase flow Navier–Stokes model. *Journal of Computational and Applied Mathematics*, 246:144–152.

Huang, N., Long, S. R., Tung, C. C., Yuen, Y., and Bliven, F. L. 1981. A unified two-parameter wave spectral model for a general sea state. *Journal of Fluid Mechanics*, 112:203–224.

Huang, R. X. 2009. *Ocean Circulation: Wind-Driven and Thermohaline Processes*. Cambridge University Press, Cambridge.

Hughes, B. 1978. The effect of internal waves on surface wind waves 2. Theoretical analysis. *Journal of Geophysical Research*, 83(C1):455–465.

Hughes, B. and Grant, H. 1978. The effect of internal waves on surface wind waves 1. Experimental measurements. *Journal of Geophysical Research*, 83(C1):443–454.

Hühnerfuss, H., Gericke, A., Alpers, W., Theis, R., Wismann, V., and Lange, P. A. 1994. Classification of sea slicks by multifrequency radar techniques: New chemical insights and their geophysical implications. *Journal of Geophysical Research*, 99(C5):9835–9845.

Hühnerfuss, H. and Walter, W. 1987. Attenuation of wind waves by monomolecular sea slicks. *Journal of Geophysical Research*, 92(C4):3961–3963.

Hwang, P. A., Poon, Y.-K., and Wu, J. 1991. Temperature effects on generation and entrainment of bubbles induced by a water jet. *Journal of Physical Oceanography*, 21:1602–1605.

Hwang, P. A., Wang, D. W., Walsh, E. J., Krabill, W. B., and Swift, R. N. 2000a. Airborne measurements of the directional wavenumber spectra of ocean surface waves. Part I: Spectral slope and dimensionless spectral coefficient. *Journal of Physical Oceanography*, 30(11):2753–2767.

Hwang, P. A., Wang, D. W., Walsh, E. J., Krabill, W. B., and Swift, R. N. 2000b. Airborne measurements of the directional wavenumber spectra of ocean surface waves. Part 2. Directional distribution. *Journal of Physical Oceanography*, 30(11):2768–2787.

Iafrati, A. 2009. Numerical study of the effects of the breaking intensity on wave breaking flows. *Journal of Fluid Mechanics*, 622:371–411.

Ivanov, A. 2000. Oil pollution of the sea on Kosmos -1870 and Almaz-1 radar imagery. *Earth Observation and Remote Sensing*, 15(6):949–966.

Irisov, V. 2014. Model of wave breaking and microwave emissivity of sea surface. *International Geoscience and Remote Sensing Symposium*. July 13–18, 2014, Quebec City, Canada. Presentation. https://www.researchgate.net/publication/269395853_MODEL_OF_WAVE_BREAKING_AND_MICROWAVE_EMISSIVITY_OF_SEA_SURFACE

Irisov, V. and Plant, W. 2016. Phillips' Lambda function: Data summary and physical model. *Geophysical Research Letters*, 43(5):2053–2058. Doi: 10.1002/2015GL067352.

Irisov, V. G. and Voronovich, A. G. 2011. Numerical Simulation of Wave Breaking. *Journal of Physical Oceanography*, 41(2):346–364. Doi: 10.1175/2010JPO4442.1.

Janssen, P. 2009. *The Interaction of Ocean Waves and Wind.* Cambridge University Press, Cambridge.

Jha, M. N., Levy, J., and Gao, Y. 2008. Advances in remote sensing for oil spill disaster management: State-of-the-art sensors technology for oil spill surveillance. *Sensors (Basel)*, 8(1):236–255.

Jiménez, J. 2015. Direct detection of linearized bursts in turbulence. *Physics of Fluids*, 27(6):065102-1–065102-14

Johnson, B. D. and Cooke, R. C. 1979. Bubble populations and spectra in coastal waters: A photographic approach. *Journal of Geophysical Research*, 84(C7):3761–3766.

Johnson, B. D. and Wangersky, P. J. 1987. Microbubbles: Stabilization by monolayers of adsorbed particles. *Journal of Geophysical Research*, 92(C13):14641–14647.

Karaev, V., Kanevsky, M., and Meshkov, E. 2008. The effect of sea surface slicks on the Doppler spectrum width of a backscattered microwave signal. *Sensors (Basel)*, 8(6):3780–3801.

Keller, W. C., Plant, W. J., and Weissman, D. E. 1985. The dependence of X band microwave sea return on atmospheric stability and sea state. *Journal of Geophysical Research*, 90(C1):1019–1029.

Kerman, B. R. 1984. Underwater sound generation by breaking wind waves. *The Journal of the Acoustical Society of America*, 75(1):149–165.

Kerry, N. J., Burt, R. J., Lane, N. M., and Bagg, M. T. 1984. Simultaneous radar observations of surface slicks and *in situ* measurements of internal waves. *Journal of Physical Oceanography*, 14(8):1419–1422.

Kinsman, B. 2012. *Wind Waves: Their Generation and Propagation on the Ocean Surface.* Dover Earth Science, New York.

Kitaigorodskii, S. A. 1973. *Physics of Air-Sea Interaction.* Israel Program of Scientific Translations, Jerusalem.

Kitaigorodskii, S. A. 1984. On the fluid dynamical theory of turbulent gas transfer across an air-sea interface in the presence of breaking wind-waves. *Journal of Physical Oceanography*, 14(5):960–972.

Klemas, V. 2010. Tracking oil slicks and predicting their trajectories using remote sensors and models: Case studies of the Sea Princess and Deepwater Horizon oil spills. *Journal of Coastal Research*, 26(5):789–797.

Knobloch, E. and Moehlis, J. 2000. Burst mechanisms in hydrodynamics. In *Nonlinear Instability, Chaos and Turbulence*, Debnath, L. and Riahi, D. (eds.). Vol. II, pp. 237–287. Computational Mechanics Publications, Southampton.

Kolobaev, P. A. 1976. Investigation of the concentration and statistical size distribution of wind produced bubbles in the near-surface ocean. *Oceanology*, 15:659–661 (translated from Russian).

Komen, G. J., Cavaleri, L., Donelan, M., Hasselmann, K., Hasselmann, S., and Janssen, P. A. E. M. 1996. *Dynamics and Modelling of Ocean Waves*. Cambridge University Press, Cambridge.

Kondratyev, K. Ya., Ivlev, L. S., Krapivin, V. F., and Varostos, C. A. 2005. *Atmospheric Aerosol Properties: Formation, Processes and Impacts*. Springer/Praxis, Chichester, UK.

Krasitskii, V. P. and Kozhelupova, N. G. 1995. On conditions for five wave resonant interactions of surface gravity waves. *Oceanology*, 34(4):435–439 (translated from Russian).

Kraus, E. B. and Businger, J. A. 1994. *Atmosphere-Ocean Interaction*. Oxford University Press, New York.

Kudryavtsev, V. N., Makin, V. K., and Chapron, B. 1999. Coupled sea surface atmosphere model: 2. Spectrum of short wind waves. *Journal of Geophysical Research*, 104(C4):7625–7639.

Lafon, C., Piazzola, J., Forget, P., Le Calve, O., and Despiau, S. 2004. Analysis of the variations of the whitecap fraction as measured in a coastal zone. *Boundary-Layer Meteorology*, 111(2):339–360.

Lamarre, E. and Melville, W. K. 1994. Void-fraction measurements and sound-speed fields in bubble plumes generated by breaking waves. *The Journal of the Acoustical Society of America*, 95:1317–1328.

Lamb, H. 1932. *Hydrodynamics*. 6th edition. Cambridge University Press, Cambridge, UK.

Lavrenov, I. 2003. *Wind-Waves in Oceans: Dynamics and Numerical Simulations*. Springer, Berlin, Heidelberg.

Leifer, I., Caulliez, G., and de Leeuw, G. 2006. Bubbles generated from wind-steepened breaking waves: 2. Bubble plumes, bubbles, and wave characteristics. *Journal of Geophysical Research*, 111(C6):C06021.

Leighton, T. G. 1994. *The Acoustic Bubble*. Academic Press, San Diego, CA.

Leikin, I. A. and Rosenberg, A. D. 1980. On the high-frequency range of the wind wave spectrum. *USSR Academy of Sciences (Doklady Akademii Nauk SSSR)*, 255:455–458.

Lewis, E. R. and Schwartz, S. E. 2004. *Sea Salt Aerosol Production: Mechanisms, Methods, Measurements, and Models—A Critical Review*. American Geophysical Union, Washington DC.

Lin, P. and Liu, P. L.-F. 1998. A numerical study of breaking waves in the surf zone. *Journal of Fluid Mechanics*, 359:239–264.

Lips, U., Lips, I., Liblik, T., and Elken, J. 2008. Estuarine transport versus vertical movement and mixing of water masses in the Gulf of Finland (Baltic Sea). In *US/EU-Baltic Symposium "Ocean Observations, Ecosystem-Based Management & Forecasting,"* Tallinn, 27–29 May, 2008. IEEE Conference Proceedings, pp. 326–333. http://www.baltex-research.eu/baltic2009/downloads/Student_presentations/and choose Bornholm_Taavi_very_final_version.pptx

Longuet-Higgins, M. S. and Turner, J. S. 1974. An "entraining plume" model of a spilling breaker. *Journal of Fluid Mechanics*, 63(1):1–20.

Lubin, P. and Glockner, S. 2015. Numerical simulations of three-dimensional plunging breaking waves: Generation and evolution of aerated vortex filaments. *Journal of Fluid Mechanics*, 767:364–393.

Ma, Q. 2010. *Advances in Numerical Simulation of Nonlinear Water Waves. Advances in Coastal and Ocean Engineering*, Vol. 11. The World Scientific Publishing Co., Singapore.

Makin, V. K. and Kudryavtsev, V. N. 1999. Coupled sea surface-atmosphere model. 1. Wind over waves coupling. *Journal of Geophysical Research*, 104(C4): 7613–7623.

Mandelbrot, B. B. 1983. *The Fractal Geometry of Nature*, 3rd edition. W. H. Freeman, New York.

Manneville, P. 2010. *Instabilities, Chaos and Turbulence (ICP Fluid Mechanics)*, 2nd edition. Imperial College Press, London.

Massel, S. R. 2007. *Ocean Waves Breaking and Marine Aerosol Fluxes*. Springer, New York.

Medwin, H. 1977. In situ acoustic measurements of microbubbles at sea. *Journal of Geophysical Research*, 82(6):971–976.

Medwin, H. and Breitz, N. D. 1989. Ambient and transient bubble spectral densities in quiescent seas and under spilling breakers. *Journal of Geophysical Research*, 94(C9):12751–12759.

Melville, W. K. 1994. Energy dissipation by breaking waves. *Journal of Physical Oceanography*, 24:2041–2049.

Melville, W. K. 1996. The role of surface-wave breaking in air-sea interaction. *Annual Review of Fluid Mechanics*, 28:279–321.

Melville, W. K. and Matusov, P. 2002. Distribution of breaking waves at the ocean surface. *Nature*, 417:58–63.

Merzi, N. and Graft, W. H. 1985. Evaluation of the drag coefficient considering the effects of mobility of the roughness elements. *Annales Geophysicae*, 3(4):473–478.

Michell, J. H. 1893. The highest waves in water. *Philosophical Magazine*, Series 5. 36(222):430–437. Doi: 10.1080/14786449308620499.

Migliaccio, M., Nunziata, F., and Buono, A. 2015. SAR polarimetry for sea oil slick observation. *International Journal of Remote Sensing*, 36(12):3243–3273.

Miropol'sky, Yu. Z. 2001. *Dynamics of Internal Gravity Waves in the Ocean*. Kluwer Academic Publishers, Dordrecht.

Mitsuyasu, H. 2002. A historical note on the study of ocean surface waves. *Journal of Oceanography*, 58:109–120.

Mitsuyasu, H. and Honda, T. 1974. The high frequency spectrum of wind generated waves. *Journal of Physical Oceanography*, 30:185–195.

Mitsuyasu, H. and Honda, T. 1982. Wind-induced growth of water waves. *Journal of Fluid Mechanics*, 123:425–442.

Mityagina, M. I., Pungin, V. G., Smirnov, A. V., and Etkin, V. S. 1991. Changes of the energy-bearing region of the sea surface wave spectrum in an internal wave field based on remote observation data. *Izvestiya, Atmospheric and Oceanic Physics*, 27(11):925–929 (translated from Russian).

Miyake, Y. and Abe, T. 1948. A study of the foaming of sea water. Part 1. *Journal of Marine Research*, 7(2):67–73.

Moiseev, S. S., Pungin, V. G., and Oraevsky, V. N. 1999. *Non-Linear Instabilities in Plasmas and Hydrodynamics*. CRC Press, Boca Raton, FL.

Moiseev, S. S. and Sagdeev, R. Z. 1986. Problems of secondary instabilities in hydro-dynamics and in plasma. *Radiophysics and Quantum Electronics*, 29(9):808–812 (translated from Russian).

Monahan, E. C. 1968. Sea spray as a function of low elevation wind speed. *Journal of Geophysical Research*, 73(4):1127–1137.

Monahan, E. C. 1969. Fresh water whitecap. *Journal of the Atmospheric Science*, 26(5):1026–1029.

Monahan, E. C. 1971. Oceanic whitecaps. *Journal of Physical Oceanography*, 1:139–144.

Monahan, E. C. and Lu, M. 1990. Acoustically relevant bubble assemblages and their dependence on meteorological parameters. *IEEE Journal of Oceanic Engineering*, 15(4):340–349.

Monahan, E. C. and MacNiocaill, G. 1986. *Oceanic Whitecaps*. D. Reidel, Dordrecht, The Netherlands.

Monahan, E. C. and O'Muircheartaigh, I. G. 1980. Optimal power-law description of oceanic whitecap coverage dependence on wind speed. *Journal of Physical Oceanography*, 10(2):2094–2099.

Monahan, E. C. and O'Muircheartaigh, I. G. 1986. Whitecaps and passive remote sensing. *International Journal of Remote Sensing*, 7(5):627–642.

Monahan, E. C., Spiel, D. E., and Davidson, K. L. 1982. Whitecap aerosol productivity deduced from simulation tank measurements. *Journal of Geophysical Research*, 87(C11):8898–8904.

Monahan, E. C. and Zietlow, C. R. 1969. Laboratory comparison of fresh-water and salt-water whitecaps. *Journal of Geophysical Research*, 74(28):6961–6966.

Monin, A. S. and Ozmidov, R. V. 1985. *Turbulence in the Ocean*. D. Reidel Publishing Company, Dordrecht, The Netherlands.

Monin, A. S. and Yaglom, A. M. 2007. *Statistical Fluid Mechanics: Mechanics of Turbulence*. Vol. 1. Dover Publications, Mineola, NY.

Mulhearn, P. J. 1981. Distribution of microbubbles in coastal water. *Journal of Geophysical Research*, 86(C7):6429–6434.

Norris, S. J., Brooks, I. M., Moat, B. I., Yelland, M. J., de Leeuw, G., Pascal, R. W., and Brooks, B. 2013. Near-surface measurements of sea spray aerosol production over whitecaps in the open ocean. *Ocean Science*, 9(1):133–145.

Onstott, R. and Rufenach, C. 1992. Shipboard active and passive microwave measurement of ocean surface slicks off the southern California coast. *Journal of Geophysical Research*, 97(C4):5315–5323.

Paget, A. C., Bourassa, M. A., and Anguelova, M. D. 2015. Comparing *in situ* and satellite-based parameterizations of oceanic whitecaps. *Journal of Geophysical Research: Oceans*, 120(4):2826–2843.

Peltzer, R. D. and Griffin, O. M. 1988. Stability of a three-dimensional foam layer in sea water. *Journal of Geophysical Research*, 93(C9):10804–10812.

Petrov, V. V. 1979a. Dynamics of nonlinear waves in a stratified ocean. *Izvestiya, Atmospheric and Oceanic Physics*, 15(7):508–513 (translated from Russian).

Petrov, V. V. 1979b. On the nonlinear damping of long surface waves in a stratified ocean. *Izvestiya, Atmospheric and Oceanic Physics*, 15(9):697–699 (translated from Russian).

Phelps, A. D. and Leighton, T. G. 1998. Oceanic bubble population measurements using a buoy-deployed combination frequency technique. *IEEE Journal of Oceanic Engineering*, 23(4):400–410.

Phillips, O. M. 1980. *The Dynamics of the Upper Ocean*. 2nd edition. Cambridge University Press, Cambridge.

Phillips, O. M. 1985. Spectral and statistical properties of the equilibrium range in wind-generated gravity waves. *Journal of Fluid Mechanics*, 156(1):505–531.

Phillips, O. M. and Banner, M. L. 1974. Wave breaking in the presence of wind drift and swell. *Journal of Fluid Mechanics*, 66(4):625–640.

Phillips, O. M. and Hasselmann, K. 1986. *Wave Dynamics and Radio Probing of the Ocean Surface*. Plenum Press, New York.

Phillips, O. M., Posner, F. L., and Hansen, J. P. 2001. High range resolution radar measurements of the speed distribution of breaking events in wind-generated ocean waves: Surface impulse and wave energy dissipation rates. *Journal of Physical Oceanography*, 31(2):450–460.

Pierson, W. J. and Moskowitz, L. 1964. A proposed spectral form for fully developed wind seas based on the similarity theory of S. A. Kitaigorodskii. *Journal of Geophysical Research*, 69(24):5181–5190.

Plant, W. J. 2015. Short wind waves on the ocean: Wavenumber-frequency spectra. *Journal of Geophysical Research*, 120(3):2147–2158.

Popov, V. A. and Chashechkin, Yu. D. 1979. On the structure of thermohaline convection in a stratified fluid. *Izvestiya, Atmospheric and Oceanic Physics*, 15(9):668–675 (translated from Russian).

Radko, T. 2013. *Double-Diffusive Convection*. Cambridge University Press, Cambridge, UK.

Rahman, M. 2005. *Instability of Flows (Advances in Fluid Mechanics)*. WIT Press Computational Mechanics, Southampton, UK.

Raizer, V. 2007. Macroscopic foam-spray models for ocean microwave radiometry. *IEEE Transactions on Geoscience and Remote Sensing*, 45(10):3138–3144. Doi: 10.1109/TGRS.2007.895981.

Raizer, V. 2008. Modeling of L-band foam emissivity and impact on surface salinity retrieval. In *Proceedings of International Geoscience and Remote Sensing Symposium*, July 6–11, Boston, MA, Vol. 4, pp. IV-930–IV-933. Doi: 10.1109/IGARSS.2008.4779876.

Raizer, V. Y. 1994. Wave spectrum and foam dynamics via remote sensing. In *Satellite Remote Sensing of the Ocean Environment*, I. S. F. Jones, Y. Sugimori, and R. W. Stewart (eds.), pp. 301–304. Seibutsu Kenkyusha, Japan.

Raizer, V. Y., Novikov, V. M., and Bocharova, T. Y. 1994. The geometrical and fractal properties of visible radiances associated with breaking waves in the ocean. *Annales Geophysicae*, 12(12):1229–1233.

Raizer, V. Yu. and Cherny, I. V. 1994. *Microwave diagnostics of ocean surface. "Mikrovolnovaia diagnostika poverkhnostnogo sloia okeana."* Gidrometeoizdat. Sankt-Peterburg. Library of Congress, LC classification (full) GC211.2 .R35 1994. (in Russian).

Raizer, V. Yu. and Novikov, V. M. 1990. Fractal dimension of ocean-breaking waves from optical data. *Izvestiya, Atmospheric and Oceanic Physics*, 26(6):491–494 (translated from Russian).

Raizer, V. Yu. and Sharkov, E. A. 1980. On the dispersed structure of sea foam. *Izvestiya, Atmospheric and Oceanic Physics*, 16(7):548–550 (translated from Russian).

Raizer, V. Yu., Smirnov, A. V., and Etkin, V. S. 1990. Dynamics of the large-scale structure of the disturbed surface of the ocean from analysis of optical images. *Izvestiya, Atmospheric and Oceanic Physics*, 26(3):199–205 (translated from Russian).

Rapp, R. J. and Melville, W. K. 1990. Laboratory experiments of deep-water breaking waves. *Philosophical Transactions of the Royal Society of London. Series A.*, 331(1622):735–800.

Reul, N. and Chapron, B. 2003. A model of sea-foam thickness distribution for passive microwave remote sensing applications. *Journal of Geophysical Research*, 108(C10):3321.

Riahi, D. N. 1996. *Mathematical Modeling and Simulation in Hydrodynamic Stability.* World Scientific Pub Co Inc., Singapore.

Romeiser, R., Alpers, W., and Wismann, V. 1997. An improved composite surface model for the radar backscattering cross section of the ocean surface. Part I: Theory of the model and optimization/validation by scatterometer data. *Journal Geophysical Research*, 102(C11):25237–25250.

Ross, D. B. and Cardone, V. 1974. Observation of oceanic whitecaps and their relation to remote measurements of surface wind speed. *Journal of Geophysical Research*, 79:444–452.

Rutkevich, P. B., Tur, A. V., and Yanovskiy, V. V. 1989. Interaction between surface and internal waves in an arbitrary stratified ocean. *Izvestiya, Atmospheric and Oceanic Physics*, 25(10):794–798 (translated from Russian).

Salberg, A.-B., Rudjord, O., and Solberg, A. H. S. 2014. Oil spill detection in hybrid-polarimetric SAR images. *IEEE Transactions on Geoscience Remote Sensing*, 52(10):6521–6533.

Salisbury, D. J., Anguelova, M. D., and Brooks, I. M. 2013. On the variability of white-cap fraction using satellite-based observations. *Journal of Geophysical Research*, 118(11):6201–6222.

Schmitt, R. W. 1994. Double diffusion in oceanography. *Annual Review of Fluid Mechanics*, 26:255–285.

Schmitt, R. W. 2003. Observational and laboratory insights into salt finger convection. *Progress in Oceanography*, 56(3-4):419–433.

Sharkov, E. A. 2007. *Breaking Ocean Waves: Geometry, Structure and Remote Sensing.* Springer, Berlin.

Slauenwhite, D. E. and B. D. Johnson, B. D. 1999. Bubble shattering: Differences in bubble formation in fresh water and seawater. *Journal of Geophysical Research*, 104(C2):3265–3275.

Slunyaev, A., Clauss, G. F., Klein, M., and Onorato, M. 2013. Simulations and experiments of short intense envelope solitons of surface water waves. *Physics of Fluids*, 25:067105-1–067105-16. Doi: 10.1063/1.4811493.

Solberg, A. H. S., Brekke, C., and Husøy, P. O. 2007. Oil spill detection in Radarsat and Envisat SAR images. *IEEE Transactions on Geoscience and Remote Sensing*, 45(3):746–755.

Soloviev, A. and Lukas, R. 2014. *The Near-Surface Layer of the Ocean: Structure, Dynamics and Applications (Atmospheric and Oceanographic Sciences Library, Vol. 48)*, 2nd edition. Springer, Dordrecht, Heidelberg.

Spiel, D. E. 1994. The sizes of jet drops produced by air bubbles bursting on sea- and fresh-water surfaces. *Tellus B*, 46(4):325–338.

Spiel, D. E. 1998. On the births of film drops from bubbles bursting on seawater surfaces. *Journal of Geophysical Research*, 103(C11):24907–24918.

Stanic, S., Caruthers, J. W., Goodman, R. R., Kennedy, E., and Brown, R. A. 2009. Attenuation measurements across surface-ship wakes and computed bubble distributions and void fractions. *IEEE Journal of Oceanic Engineering*, 34(1):83–92.

Steele, J. H., Thorpe, S. A., and Turekian, K. K. 2010. *Elements of Physical Oceanography: A Derivative of the Encyclopedia of Ocean Sciences.* Academic Press, London.

Stramska, M. and Petelski, T. 2003. Observations of oceanic whitecaps in the north polar waters of the Atlantic. *Journal of Geophysical Research*, 108(C3):3086.

Su, M.-Y. 1987. Deep-water wave breaking: Experiments and field measurements. In *Nonlinear Wave Interactions in Fluids. The Winter Annual Meeting of the American Society of Mechanical Engineers*. The American Society of Mechanical Engineers, pp. 23–36.

Su, M.-Y. and Green, A. W. 1984. Coupled two- and three-dimensional instabilities of surface gravity waves. *Physics of Fluids*, 27(1):2595–2597.

Sugihara, Y., Tsumori, H., Ohga, T., Yoshioka, H., and Serizawa, S. 2007. Variation of whitecap coverage with wave-field conditions. *The Journal of Marine Systems*, 66(1):47–60.

Tang, C. C. H. 1974. The effect of droplets in the air-sea transition zone on the sea brightness temperature. *Journal of Physical Oceanography*, 4:579–593.

Tang, S. and Wu, J. 1992. Suppression of wind-generated ripples by natural films: A laboratory study. *Journal of Geophysical Research*, 97(C4):5301–5306.

Taylor, J. R. 1993. Anisotropy of salt fingers. *Journal of Physical Oceanography*, 23(3):554–565.

Taylor, J. R. and Buchens, P. 1989. Laboratory experiments on the structure of salt fingers. *Deep-Sea Research*, 36:1675–1704.

Taylor, J. R. and Veronis, G. 1996. Experiments on double-diffusive sugar-salt fingers at high stability ratio. *Journal of Fluid Mechanics*, 321:315–333.

Terray, E. A., Donelan, M. A., Agrawal, Y. C., Drennan, W. M., Kahma, K. K., Williams III, A. J., Hwang, P. A., and Kitaigorodskii, S. A. 1996. Estimates of kinetic energy dissipation under breaking waves. *Journal of Physical Oceanography* 26:792–807.

Thompson, D. R., Gotwols, B. L., and Sterner II, R. E. 1988. A comparison of measured surface wave spectral modulations with predictions from a wave-current interaction model. *Journal of Geophysical Research*, 93(C10):12339–12343.

Thomson, J. and Jessup, A. T. 2009. A Fourier-based method for the distribution of breaking crests from video observations. *Journal of Atmospheric and Oceanic Technology*, 26, 1663–1671.

Thorpe, S. A. 1984. The effect of Langmuir circulation on the distribution of submerged bubbles caused by breaking wind waves. *Journal of Fluid Mechanics*, 14:151–170.

Thorpe, S. A. 1995. Dynamical processes of transfer at the sea surface. *Progress in Oceanography*, 35(4):315–352.

Thorpe, S. A. 2005. *The Turbulent Ocean*. Cambridge University Press, Cambridge.

Thorpe, S. A., Bowyer, P., and Woolf, D. K. 1992. Some factors affecting the size distributions of oceanic bubbles. *Journal of Physical Oceanography*, 22(4):382–389.

Toba, Y. and Mitsuyasu, H. (eds.) 2010. *The Ocean Surface: Wave Breaking, Turbulent Mixing and Radio Probing*. Softcover reprint of hardcover first 1985 edition. Springer, The Netherlands.

Trevorrow, M. V., Vagle, S., and Farmer, D. M. 1994. Acoustical measurements of microbubbles within ship wakes. *The Journal of the Acoustical Society of America*, 95(4):1922–1930.

Turner, J. S. 1973. *Buoyancy Effects in Fluids*. Cambridge University Press, Cambridge.

Turner, J. S. 1974. Double-diffusive phenomena. *Annual Review of Fluid Mechanics*, 6:37–54.

Turner, J. S. 1978. Double-diffusive intrusions into a density gradient. *Journal of Geophysical Research*, 83(C6):2887–2901.

Vagle, S. and Farmer, D. M. 1992. The measurement of bubble-size distributions by acoustical backscatter. *Journal of Atmospheric and Oceanic Technology*, 9(5):630–644.

Veron, F. 2015. Ocean Spray. *Annual Review of Fluid Mechanics*, 47:507–538.

Veron, F., Hopkins, C., Harrison, E. L., and Mueller, J. A. 2012. Sea spray spume droplet production in high wind speeds. *Geophysical Research Letters*, 39:L16602. Doi: 10.1029/2012GL052603.

Voliak, K. I. 2002. *Selected Papers. Nonlinear Waves in the Ocean*. Nauka, Moscow (in Russian and English).

Volyak, K. I., Grushin, V. A., Ivanov, A. V., Lyakhov, G. A., and Shugan, I. V. 1985. Interaction of randomly modulated surface waves. *Izvestiya, Atmospheric and Oceanic Physics*, 21(11):895–901 (translated from Russian).

Volyak, K. I., Lyakhov, G. A., and Shugan, I. V. 1987. Surface wave interaction. Theory and capability of oceanic remote sensing. In *Oceanic Remote Sensing*. F. V. Bunkin and K. I. Volyak (eds.), Nova Science Publishers, Commack, New York, pp. 107–145. (translated from Russian).

Walsh, A. L. and Mulhearn, P. J. 1987. Photographic measurements of bubble populations from breaking wind waves at sea. *Journal of Geophysical Research*, 92:14553–14565.

Weaire, D. and Hutzler, S. 1999. *The Physics of Foams*. Clarendon Press, Oxford.

Wei, Y. and Wu, J. 1992. In situ measurements of surface tension, wave damping, and wind properties modified by natural films. *Journal of Geophysical Research*, 97(C4):5307–5313.

Woolf, D. K. 2001. Bubbles. In *Encyclopedia of Ocean Sciences*, pp. 352–357. Academic Press, San Diego.

Wu, J. 1979. Spray in atmospheric surface layer: Review and analysis of laboratory and oceanic results. *Journal of Geophysical Research*, 84(C4):1693–1704.

Wu, J. 1988a. Bubbles in the near-surface ocean: A general description. *Journal of Geophysical Research*, 93(C1):587–590.

Wu, J. 1988b. Variations of whitecap coverage with wind stress and water temperature. *Journal of Physical Oceanography*, 18(10):1448–1453.

Wu, J. 1989a. Contributions of film and jet drops to marine aerosols produced at the sea surface. *Tellus B*, 41(4):469–473.

Wu, J. 1989b. Suppression of oceanic ripples by surfactant—spectral effects deduced from sun glitter, wave-staff and microwave measurements. *Journal of Physical Oceanography*, 19:238–245.

Wu, J. 1990a. On parametrization of sea spray. *Journal of Geophysical Research*, 95(C10):18269–18279.

Wu, J. 1990b. Vertical distribution of spray droplets near the sea surface: Influence of jet drop ejection and surface tearing. *Journal of Geophysical Research*, 95(C6):9775–9778.

Wu, J. 1992a. Bubble flux and marine aerosol spectra under various wind velocities. *Journal of Geophysical Research*, 97(C2):2327–2333.

Wu, J. 1992b. Individual characteristics of whitecaps and volumetric description of bubbles. *IEEE Journal of Oceanic Engineering*, 17(1):150–158.

Wu, J. 1993. Production of spume drops by the wind tearing of wave crests: The search for quantification. *Journal of Geophysical Research*, 98(C10):18221–18227.

Wu, J. 2000. Bubbles produced by breaking waves in fresh and salt waters. *Journal of Physical Oceanography*, 30:1809–1813.

Yaglom, A. M. and Frisch, U. 2012. *Hydrodynamic Instability and Transition to Turbulence. 100 (Fluid Mechanics and Its Applications)*. Springer, Dordrecht, Heidelberg.

Young, I. R. 1999. *Wind Generated Ocean Waves*. Elsevier, Oxford, UK.

Yuen, H. C. and Lake, B. M. 1982. Nonlinear dynamics of deep-water gravity waves. In *Advances in Applied Mechanics*, Chia-Shun Yih (ed.), Vol. 22, pp. 67–229. Academic Press, New York.

Zakharov, V. E. 1968. Stability of periodic waves of finite amplitude on the surface of a deep water. *Journal of Applied Mechanics and Technical Physics*, 9(2):190–194 (translated from Russian).

Zakharov, V. E. and Zaslavskii, M. M. 1982. The kinetic equation and Kolmogorov spectra in the weak turbulence theory of wind waves. *Izvestiya, Atmospheric and Oceanic Physics*, 18(9):747–753 (translated from Russian).

Zaslavskiy, M. M. 1996. On the role of four-wave interactions in formation of space–time spectrum of surface waves. *Izvestiya, Atmospheric and Oceanic Physics*, 31(4):522–528 (translated from Russian).

Zhang, B., Perrie, W., Li, X., and Pichel, W. G. 2011. Mapping sea surface oil slicks using Radarsat-2 Quad-polarization SAR image. *Geophysical Research Letters*, 38(10):1–5.

Zhao, D. and Toba, Y. 2001. Dependence of whitecap coverage on wind and wind-wave properties. *Journal of Oceanography*, 57(5):603–616.

Zheng, Q. A., Klemas, V., and Hsu, Y.-H. L. 1983. Laboratory measurement of water surface bubble life time. *Journal of Geophysical Research*, 88(C1):701–706.

Zhurbas, V. M. and Ozmidov, R. V. 1983. Formation of stepped finestructure in the ocean by thermohaline intrusions. *Izvestiya Atmosphere and Oceanic Physics*, 19(12):977–982 (translated from Russian).

Zhurbas, V. M. and Ozmidov, R. V. 1984. Forms of step-like structures of the oceanic thermocline and their generation mechanism. *Oceanology-USSR*, 24(2):153–157.

Zhurbas, V. M. *Lecture on Oceanography*. http://msi.ttu.ee/~elken/Zhurbas_L08.pdf

Zitha, P., Banhart, J., and Verbist, G. 2000. *Foams, Emulsions and Their Applications*. MIT-Verlag, Bremen, Germany.

3

Microwave Emission of the Ocean

3.1 Main Factors and Mechanisms

3.1.1 Introduction

Microwave remote sensing is based on the measurements and analyses of the thermal emission of environmental media and/or natural objects. Microwaves are defined as the part of the electromagnetic spectrum in the range of wavelengths $\lambda = 0.1$–100 cm or in the range of frequencies 300–0.3 GHz. The microwave frequency bands are specified as several standard intervals:

P band	0.230–1.000 GHz	
UHF band	430–1300 MHz	
L band	1.530–2.700 GHz	
S band	2.700–3.500 GHz	
C band	3.700–4.200 GHz	(Downlink)
	5.925–6.425 GHz	(Uplink)
X band	7.250–7.745 GHz	(Downlink)
	7.900–8.395 GHz	(Uplink)
Ku band	10.7–18.0 GHz	(has multiple acceptations)
Ka band	18.0–40.0 GHz	(has multiple acceptations)
V band	40–75 GHz	
W band	75–110 GHz	
F band	90–140 GHz	(waveguide specifications)
D band	110–170 GHz	
G band	140–300 GHz	

Microwaves are highly sensitive to variations of structure parameters of environmental media that is associated with a larger depth of penetration in comparison to infrared and optical electromagnetic waves. Along with that,

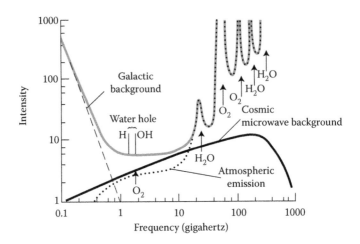

FIGURE 3.1
Intensity spectrum in degrees Kelvin of galaxy and atmospheric emission. (http://pages. uoregon.edu/jimbrau/BrauImNew/Chap28/6th/28_17Figure-F.jpg.)

some microwaves can pass through the atmosphere and clouds and there- fore, they are used for remote sensing of Earth's surface. Microwave sensors also have all-weather capability, which is their principal advantage.

Figure 3.1 illustrates the electromagnetic spectrum and atmospheric windows for microwave background radiation in terms of the emission intensity. The microwave emission of the sky is defined by galactic back- ground, cosmic microwave background, and atmospheric emission. At the L band, total microwave contributions from the sky and the atmosphere have minimal and relatively stable level that provides great possibilities for direct observations of the ocean surface. The entire band of wavelengths around $\lambda = 21$ cm (frequency of 1.420 GHz) is an interesting part of the electromag- netic spectrum and is known as the water hole. It is spin-flip line of hydrogen because hydrogen is the simplest atom in the universe. Thus, $\lambda = 21$ cm is the wavelength of simple emission of hydrogen.

The effectiveness and reliability of microwave diagnostics of the ocean depend to a large extent on the knowledge of physical mechanisms and characteristics of thermal microwave emission. The scope of the task includes practically all major problems of electromagnetic wave propagation, statistical radiophysics, and signal processing. Generally speaking, the theory of microwave diagnos- tics is based on the solutions of Maxwell's equations for a random onstation- ary lossy dielectric medium, including both multiscale surface and volume nonuniformities. Such a multiple electromagnetic task cannot be completed without the corresponding *computational and numerical resources*.

Many aspects of microwave remote sensing theory and practice are pro- pounded in several books (Basharinov et al. 1974; Bogorodskiy et al. 1977; Ulaby et al. 1981, 1982, 1986; Tsang et al. 1985; Shutko 1986; Scou 1989; Janssen

1993; Fung 1994; Sharkov 2003; Joseph 2005; Woodhouse 2005; Matzler 2006; Fung and Chen 2010; Robinson 2010; Ulaby and Long 2013; Martin 2014; Njoku 2014; Grankov and Milshin 2015; Lavender and Lavender 2015).

However, as follows from these and many other literature sources, the capabilities of microwave radiometry and imagery to observe ocean dynamic features and disturbances have not been fully realized. Along with the topics, an overall concept of ocean microwave diagnostics was missing until now. Actually, it is still not known for certain what types of hydrodynamic processes and/or events in the ocean are potentially observable by a microwave radiometer and what are not observable at all or might be detected somehow. The ocean microwave data existing at the moment just provide a guidance but do not answer this question.

One possible reason is lack of understanding how to measure and/or investigate multiscale highly dynamic processes using passive microwave techniques. Another reason is the absence of good evidence for believing that high-resolution multifrequency polarimetric radiometer-imager can offer more useful information than one-frequency regular microwave radar. The follow-up discussion and the material presented may clear up this misunderstanding.

3.1.2 Microwave Characterization

Figure 3.2 illustrates the basic elements of the microwave remote sensing model of the ocean environment. The primary attribute is hydrodynamic

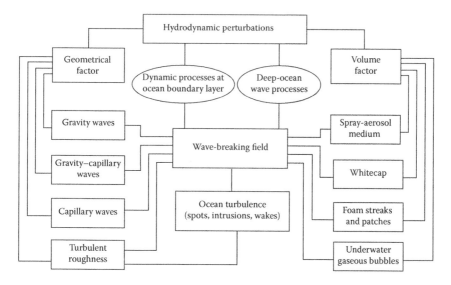

FIGURE 3.2
Elements of the radio-hydro-physical model of the ocean surface. (Undated from Cherny I. V. and Raizer V. Yu. *Passive Microwave Remote Sensing of Oceans.* 195 p. 1998. Copyright Wiley-VCH Verlag GmbH & Co. KGaA. Reproduced with permission.)

perturbations acting on the air–sea interface that yield measurable micro-wave response. While the ocean is a multicomponent system, two main classes—geometrical and volume perturbations—should be considered and taken into account in microwave studies.

The geometrical class relates to the ocean surface waves whose geometry and statistics are defined by interactions between oceans and the atmosphere. In general, surface waves are presented as a nonstationary and nonuniform field of multiscale surface disturbances, which have both deterministic and random components. The class of geometric perturbations includes a number of subclasses related to gravity waves, gravity–capillary waves, capillary waves, and turbulent roughness as well. Wave–wave interactions and the strong intermittent behavior of wave components may trigger changes in ocean microwave emission, depending on environmental conditions or situations.

The physical mechanism of ocean microwave emission associated with the geometrical factor includes the following principal effects:

- Mirror reflection from a small-scale roughness surface
- Diffused incoherent scattering on multiscale surface irregularities
- Coherent scattering from correlated surface irregularities
- Resonance scattering from surface irregularities with geometrical sizes that are comparable with the electromagnetic wavelength
- Multiple scattering and shadows on large-scale irregularities

The volume class of nonuniformities represents a number of two-phased (air–water) disperse systems, which are foam, whitecap, bubble populations, sea spray, droplets, or their aggregates. These highly dynamic inhomogeneous natural objects are formed on the air–sea interface as a result of waves breaking and aeration processes, migration of deep-ocean gaseous bubbles, cavitating flows, or others causes.

It is also important to remember that the electromagnetic properties of natural oceanic disperse media and their contributions to ocean microwave emission are quite different. This statement has been established in the late 1970s and explained in detail in two books (Raizer and Cherny 1994; Cherny and Raizer 1998). The main electromagnetic mechanisms here are single and multiple scattering, absorption, and extinction, including resonance (Mie, Rayleigh) and cooperative radiation effects occurring in polydisperse systems of closely packed particles (bubbles, droplets).

The elements and relationships shown in Figure 3.2 may be ambiguous and must be specified and adjusted as knowledge is acquired about ocean hydrodynamic and wave propagation phenomena. Because the inverse problem of remote sensing is, *a priori*, incorrect mathematically, the solution requires complementary information about studied processes or phenomena. Such information is usually obtained using *in situ* measurements. In this case, flexible algorithms and numerical approximations are able to provide a reliable and comprehensive analysis of ocean emissivity.

Introducing different factors into the model (Figure 3.2) can be made consecutively in accordance with the chosen conditions but not randomly. Geometrical nonuniformities are an integral part of the ocean–atmosphere system. Disperse media can be incorporated only in certain conditions, for example, in cases of high wind and gales. In fact, there are no universally accepted methods for the description of the ocean environment and related electromagnetic problems. Therefore, the creation of a universal microwave electromagnetic model of the ocean–atmosphere system is an extremely difficult task because a number of key parameters and factors should be specified and involved to provide adequate characterization of environments.

However, it is possible to consider and investigate microwave impacts from each factor, taken separately, at least in context with the existing experimental and theoretical data. Then, we combine all of them into a unit model. Thus, we come to the problem of multifactor and multiparameter description offering numerical modeling and simulation of different microwave ocean scenes and/or scenarios with a large set of hydrodynamic variables.

3.1.3 Basic Relationships

In accordance with the Rayleigh–Jeans approximation of Planck's law, the intensity of intrinsic microwave radiation is expressed in terms of brightness temperature T_B, which is a product of the coefficient of emission (emissivity) κ and the thermodynamic (physical) temperature T_0:

$$T_B = \kappa T_0. \tag{3.1}$$

In the simplest case of a smooth surface, the coefficient of emission is defined through the complex Fresnel reflection coefficient:

$$\kappa_{h,v} = 1 - |r_{h,v}|^2, \tag{3.2}$$

$$r_h = \frac{\cos\theta - \sqrt{\varepsilon - \sin^2\theta}}{\cos\theta + \sqrt{\varepsilon - \sin^2\theta}}, \tag{3.3}$$

$$r_v = \frac{\varepsilon\cos\theta - \sqrt{\varepsilon - \sin^2\theta}}{\varepsilon\cos\theta + \sqrt{\varepsilon - \sin^2\theta}}, \tag{3.4}$$

where $\kappa_{h,v}$ and $r_{h,v}$ are emission and reflection coefficients for horizontal (index "h") and vertical (index "v") polarizations, respectively; $\varepsilon = \varepsilon' - i\varepsilon''$ is the complex dielectric constant of the medium; and θ is an observation angle (incidence). At the nadir view angle $\theta = 0$, complex reflection coefficients $r_v = -r_h$. The Fresnel equations have some mathematical features which are investigated the best in several books (Stratton 1941; Born and Wolf 1999).

The surface brightness temperature is given by

$$T_{Bh,v} = \kappa_{h,v} T_0 = (1 - |r_{h,v}|^2) T_0, \qquad (3.5)$$

where $T_0 = \text{const}$ (in Kelvin). Relationship (3.2) represents Kirchhoff's law, which states that for thermal equilibrium for a particular surface, the monochromatic emissivity equals the monochromatic absorptivity (Kirchhoff 1860; Planck 1914; Robitaille 2009).

The brightness temperature of the sea surface $T_{Bh,v}(\lambda, \theta; t, s)$ is a function of incidence angle (θ), polarization (h, v), electromagnetic wavelength (λ), temperature (t), and salinity (s) of water. This relationship is determined by the dependency of the complex permittivity of water on frequency (called "dielectric dispersion") and also its dependency on temperature and salinity $\varepsilon = \varepsilon_w(\lambda; t, s)$ (Section 3.1.4). Such a parameterization is commonly used for theoretical predictions of basic spectral and polarization dependencies of the sea surface brightness temperature (Figure 3.3).

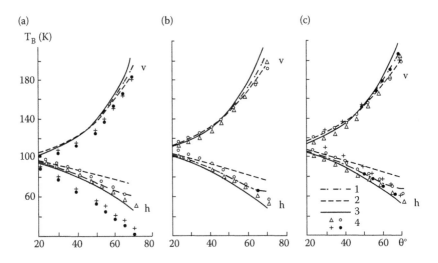

FIGURE 3.3

Dependencies of the sea surface brightness temperature on view angle. v—vertical polarization; h—horizontal polarization. Wavelength (frequency) of emission: (a) $\lambda = 21.4$ cm (1.4 GHz); (b) $\lambda = 7.5$ cm (4.0 GHz); (c) $\lambda = 4$ cm (7.5 GHz). Calculation: (1) flat water surface; the Kirchhoff model; (2) r.m.s. surface wave's slope 10°; (3) r.m.s. wave's slope 15°; (4) experimental data. (Adapted from Wu, S. T. and Fung, A. K. 1972. *Journal of Geophysical Research.* 77(30):5917–5929. Doi: 10.1029/JC077i030p05917; Cherny I. V. and Raizer V. Yu. *Passive Microwave Remote Sensing of Oceans.* 195 p. 1998. Copyright Wiley-VCH Verlag GmbH & Co. KGaA. Reproduced with permission.)

However, the fundamental Fresnel-based formulation does not provide a complete characterization of ocean thermal microwave emission. In the real world, we often observe short-term fluctuations, deviations, and trends in the brightness temperature that is explained mostly by geometric and structural perturbations (variations) of the air–sea interface. Environmental changes occur under the influence of many factors, including the ocean–atmosphere interactions, wind actions, and wave motions. Generally speaking, classical Fresnel Equations 3.3 and 3.4 may describe only mean, globally (planetary) averaged value of the brightness temperature, ignoring the influence of local irregularities.

More adequately, an extended formulation is based on the knowledge of the bistatic surface scattering coefficient. At the selected wavelength λ and polarization p, the coefficient of emission is defined as the following (Peake, 1959):

$$\kappa_p(\lambda;\theta_0,\phi_0) = 1 - \frac{1}{4\pi}\iint [\sigma_{pp}(\lambda;\theta_0,\varphi_0;\theta_s,\phi_s) + \sigma_{pq}(\lambda;\theta_0,\varphi_0;\theta_s,\varphi_s)]d\Omega_s, \quad (3.6)$$

where σ_{pp} and σ_{pq} are the bistatic scattering coefficients at co- and cross-polarizations (p = h, v) or (q = v, h); θ_0, $\varphi_0;\theta_s$, ϕ_s are the angular coordinates for incident (emitted) and scattered radiation; and $d\Omega_s = \sin\theta_s d\theta_s d\phi_s$ is the elementary solid angle.

In the case of a smooth surface (cross-polarization term $\sigma_{pq} = 0$), the scattering coefficient at horizontal or vertical polarization is

$$\sigma_{pp}(\theta_0,\phi_0;\theta_s,\phi_s) = \frac{4\pi}{\sin\theta_s}|r_p(\theta_0)|^2 \delta(\theta_s - \theta_0)\delta(\phi_s - \phi_0) \quad (3.7)$$

and formula (3.6) results in Equation 3.2 ($\theta_0 = \theta;\phi_0 = 0;p = h, v$).

Relationship (3.6) provides calculations of emissivity depending on the configuration of the air–sea interface, including both surface (geometrical) and volume nonuniformities. In this case, the total scattering coefficient can be written as the sum $\sigma_\Sigma = \sigma_{pp} + \sigma_{pq} = \left(\sigma_{pp}^{sur} + \sigma_{pp}^{vol}\right) + \left(\sigma_{pq}^{sur} + \sigma_{pq}^{vol}\right)$, where $\sigma_{pp,pq}^{sur}$ and $\sigma_{pp,pq}^{vol}$ are the terms related to surface scattering and volume scattering, respectively. The dependence of the scattering coefficients on the dielectric permittivity as a function of physical parameters remains.

A general approach for computing emissivity is based on the macroscopic theory of thermal electromagnetic fluctuations and the fluctuation dissipation theorem for distributed systems (Levin and Rytov 1973; Landau and Lifshitz 1984; Rytov et al. 1989). This rigorous electromagnetic theory describes thermal radiation from nonisothermal and nonuniform media, for example, multilayered dielectric structure with a vertical profile of the temperature (Tsang et al. 1975, 1985). For ocean microwave studies, this theory has limited application.

3.1.4 Stokes Parameters and Elements of Polarimetry

An important issue is the evaluation of the Stokes parameters (introduced by Sir George Stokes in 1852), which characterize the polarization state of a partially polarized thermal microwave emission. The modified Stokes vector in brightness temperature is

$$\bar{T}_B = \frac{\lambda^2}{k_B \cdot \eta \cdot B} \begin{bmatrix} T_v \\ T_h \\ T_3 \\ T_4 \end{bmatrix} = \begin{bmatrix} T_v \\ T_h \\ T_{45} - T_{-45} \\ T_{cl} - T_{cr} \end{bmatrix} = \begin{bmatrix} \langle |\vec{E}_v|^2 \rangle \\ \langle |\vec{E}_h|^2 \rangle \\ 2\,\mathrm{Re}\langle \vec{E}_v \vec{E}_h^* \rangle \\ 2\,\mathrm{Im}\langle \vec{E}_v \vec{E}_h^* \rangle \end{bmatrix}, \tag{3.8}$$

where $k_B = 1.38 \times 10^{-23}$ J/K is the Boltzmann's constant; η is the wave impedance of the medium; B is the bandwidth; and $\vec{E}_v$ and $\vec{E}_h$ are the emitted electric fields for vertical and horizontal polarization, respectively.

The first and second parameters T_v and T_h of the Stokes vector correspond to the brightness temperature for vertical and horizontal polarizations, respectively. The third and fourth parameters equal to $T_3 = T_{45} - T_{-45}$ and $T_4 = T_{cl} - T_{cr}$, where T_{45}, T_{-45}, T_{cl}, and T_{cr} refer to $+45°$ linear, $-45°$ linear, left-handed circularly, and right-handed circularly polarized brightness temperatures, respectively.

Numerous studies were conducted to explore the brightness temperature Stokes parameters, especially the third and fourth parameters of emission. As a result, the following geophysical approximation of the Stokes parameters for a wind-generated sea surface was established:

$$\begin{aligned} T_v &\approx T_{v0} + T_{v1}\cos\varphi + T_{v2}\cos 2\varphi \\ T_h &\approx T_{h0} + T_{h1}\cos\varphi + T_{h2}\cos 2\varphi \\ T_3 &\approx U_1 \sin\varphi + U_2 \sin 2\varphi \\ T_4 &\approx V_1 \sin\varphi + V_2 \sin 2\varphi. \end{aligned} \tag{3.9}$$

Here, U_1, U_2, V_1, V_2 are coefficients and $\varphi = \varphi_w - \varphi_0$ is the relative azimuth angle corresponding to the wind φ_w and observation φ_0 angular directions, respectively. It is assumed that coefficients of all harmonics in Equation 3.9 are functions of wind speed, incidence angle, and frequency.

The Stokes parameters are used in polarimetric airborne and spaceborne measurements of the sea surface wind vector, beginning with some pioneering works (Etkin et al. 1991; Dzura et al. 1992). Later, relationships (3.8) and (3.9) were investigated experimentally in several aircraft experiments using passive microwave polarimetric radiometers at centimeter and decimeter wavelengths.

More detailed information about the Stokes parameters, polarimetric technique, and measurements can be found in several references (Johnson et al.

1993, 1994; Yueh 1997; Yueh et al. 1995, 1997; Ruf 1998; Skou and Laursen 1998; Piepmeier and Gasiewski 2001; Lahtinen et al. 2003a,b; Piepmeier et al. 2008; Le Vine and Utku 2009).

3.1.5 Antenna and Radiometer Parameters

In the real-world, the microwave radiometer measures not the actual brightness temperature but the so-called antenna temperature, which is defined as

$$T_A = \frac{\iint\limits_{4\pi} T_B(\theta,\varphi)G_0(\theta,\varphi)d\Omega}{\iint\limits_{4\pi} G_0(\theta,\varphi)d\Omega}, \tag{3.10}$$

where $G_0(\theta, \varphi)$ is the antenna power gain function. Integral (3.10) provides a spatial average of the actual brightness temperature by the antenna main beam depending on the observation geometry.

The radiometric system consists of three main elements: receiver, transmission line, and antenna (Figure 3.4). The overall noise temperature T_s of the system and the output measured noise power P_s are

$$T_s = \eta_A T_A + (1-\eta_A)T_P + (L-1)T_P + LT_R, \quad P_s = k_B T_s B, \tag{3.11}$$

where η_A losses in antenna ($\eta_A < 1$), L is the loss factor of the transmission line, T_R is the equivalent noise temperature generated by the receiver, T_A is the antenna noise temperature, T_P is the temperature of the antenna and transmission line, and B is the filter bandwidth.

In order to obtain the value of actual brightness temperature T_B (which can be compared with theoretical data), it is necessary to complete several operations: (1) measure P_s with the highest sensitivity; (2) estimate T_s from

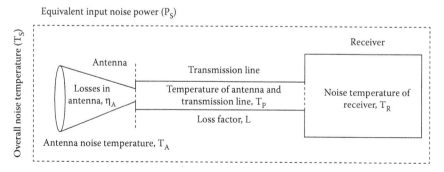

FIGURE 3.4
Basic elements of passive microwave radiometric system.

P_s that requires accurate and precision calibration of the system; (3) estimate T_A from T_s using Equation 3.11; and (4) compute T_B from Equation 3.10, which requires a detailed knowledge of antenna characteristics and observation parameters. The implementation of these procedures usually includes a special investigation of technical parameters and calibration of the antenna array system as well.

3.2 Dielectric Properties of Seawater

3.2.1 Introduction

The water molecule is a polar molecule; it has one side that is positively charged and one side that is negatively charged. The molecule is made up of two hydrogen atoms and one oxygen atom. Water structures can vary from a single molecule to clusters of hundreds of molecules bonded together. The freedom of water molecule rotation in the electric field is measured by the relaxation time (i.e., time of readjustment of molecules to equilibrium). Therefore, most authors describe the dielectric properties of seawater at microwave frequencies using a molecular theory of dielectric relaxation.

A relaxation theory of the dielectric constant of polar liquids was established in classical works (Debye 1929; Cole and Cole 1941, 1942; Von Hippel 1995). On this basis, a number of numerical models and approximations have been developed during the past several years by many authors (Hasted 1961; Stogryn 1971; Ray 1972; Rayzer et al. 1975; Klein and Swift 1977; Swift and MacIntosh 1983; Shutko 1985, 1986; Liebe et al. 1991; Meissner and Wentz 2004; Somaraju and Trumpf 2006) in order to calculate the complex dielectric constant (permittivity) of pure water, salt water, and aqueous NaCl solutions.

Some collected experimental data (Ho and Hall 1973; Akhadov 1980; Nörtemann et al. 1997; Ellison et al. 1998; Guillou et al. 1998; Ellison et al. 2003; Lang et al. 2003, 2016; Sharkov 2003; Gadani et al. 2012; Joshi and Kurtadikar 2013) demonstrate a good agreement with the theory (or with the suggested approximations) but some of them do not. The discussion of these studies and results is beyond the scope of this book.

Meanwhile, simple estimates by formula (3.5) show that considerable errors (up to 10% at selected microwave frequencies) occur in computing the sea surface brightness temperature because of the differences between the existing numerical approximations of the water dielectric constant $\varepsilon_w(\lambda, t, s)$. It is important to note that the dielectric characteristics of natural seawater as a function of temperate and salinity are not fully investigated in a wide range of microwave frequencies that is still an issue in ocean remote sensing applications.

3.2.2 Relaxation Models

A precise knowledge of the complex dielectric constant of seawater is necessary for modeling and interpretation of ocean microwave data. In microwave radiometry, the following dielectric models of water and salt solutions are used:

1. The Debye equation (Debye 1929)

$$\varepsilon_w = \varepsilon'_w - i\varepsilon''_w = \varepsilon_\infty + \frac{\varepsilon_s - \varepsilon_\infty}{1 + i\omega\tau} - i\frac{\sigma}{\omega\varepsilon_0}, \tag{3.12}$$

where $\omega = 2\pi f$ is the radian frequency (in rad/s) and f is the frequency (in GHz), ε_s is the static (low frequency) permittivity, ε_∞ is the high-frequency permittivity, τ is the relaxation time (in s), σ is the ionic conductivity (in S/m), and $\varepsilon_0 = 8.854\ldots \times 10^{-12}$ F·m⁻¹ is the vacuum permittivity (electric constant). The last term in Equation 3.12 can be recalculated as $i(\sigma/\omega\varepsilon_0) = i60\sigma\lambda$. The simplicity of the Debye relaxation model is deceptive because all parameters ε_s, ε_∞, τ, and σ are functions of the temperature (t) and the salinity (s) of water as mentioned above.

2. The Cole–Cole equation (Cole and Cole 1941, 1942)

$$\varepsilon_w = \varepsilon'_w - i\varepsilon''_w = \varepsilon_\infty + \frac{\varepsilon_s - \varepsilon_\infty}{1 + (i\omega\tau)^{1-\alpha}} - i\frac{\sigma}{\omega\varepsilon_0}, \tag{3.13}$$

where α is an empirical parameter that describes the distribution of relaxation times (usually $\alpha = 0.01–0.30$). At $\alpha = 0$, the Debye Equation 3.12 is recovered from Equation 3.13. The Cole–Cole model is commonly applied at present and gives a good approximation for the complex dielectric constant of salt water. There are some differences in the parameterizations $\varepsilon_s(t, s)$, $\varepsilon_\infty(t, s)$, $\tau(t, s)$, and $\sigma(t, s)$ suggested by authors (Stogryn 1971; Klein and Swift 1977; Meissner and Wentz 2004). They do not always describe equally and adequately the dielectric dispersion $\varepsilon_w(f)$ or $\varepsilon_w(\lambda)$ of salt water. In particular, the dependencies $\varepsilon_w(t, s)$ at C, S, and L bands vary due to the inconsistency in ionic conductivity $\sigma(t, s)$.

3. The Havriliak–Negami equation (Havriliak and Negami 1967)

$$\varepsilon_w = \varepsilon'_w - i\varepsilon''_w = \varepsilon_\infty + \frac{\varepsilon_s - \varepsilon_\infty}{[1 + (i\omega\tau)^{1-\alpha}]^\beta} - i\frac{\sigma}{\omega\varepsilon_0}. \tag{3.14}$$

This is an extended dielectric model that operates with dual-parameter (α, β) distribution of the relaxation time. The exponents

(α, β) describe the asymmetry and broadness of the dielectric spectrum $\varepsilon_w(\lambda)$. The case $\alpha = 0$, $\beta = 1$ corresponds to the Debye equation; the case $\alpha \neq 0$, $\beta = 1$ gives the Cole–Cole equation; and the case $\alpha = 0$, $\beta \neq 1$ corresponds to the Cole–Davidson formula (Davidson and Cole 1951).

The Havriliak–Negami equation is used in the dielectric spectroscopy of liquid composite materials and polymers, and biological system (Kremer and Schönhals 2003; Raicu and Feldman 2015). In ocean microwave studies, this flexible multiparameter relaxation model can be applied for the dielectric characterization of organic and nonorganic seawater compounds and emulsions as well as for the description of liquid turbulent intrusions of variable density.

3.2.3 Effects of Temperature and Salinity

As mentioned above, sea surface temperature (SST) and salinity (SSS) are the two main physical parameters that should be taken into account at ocean microwave studies. The effects of SST and SSS on the complex permittivity of water are shown in Figure 3.5. These calculations were made using the Debye model and Stogryn's approximations (Stogryn 1971). Another numerical example is presented in Figure 3.6. This diagram is created using the Cola-Cola model (3.14). To obtain detailed Cola-Cola dependencies $\varepsilon_w''(\lambda)$ versus $\varepsilon_w'(\lambda)$, the electromagnetic wavelength is changed quietly from $\lambda = 0.3$ to 30 cm with very small intervals $\Delta\lambda = 0.1$ cm. From these data, it follows that the effects of SST are pronounced mostly at K and X bands, whereas the effects of SSS appear at C, S, and L bands. More detailed considerations also reveal the dependency of the complex permittivity on the relaxation time, especially in the case of salt water solutions.

The formation of microwave emission is defined by the dielectric properties of the skin layer of a media. The thickness of the skin layer is

$$\ell = \left(\frac{2\pi}{\lambda}\right)^{-1} \left\{ \frac{\varepsilon'}{2}[(1+tg^2\delta)^{1/2} - 1] \right\}^{-1/2}, \tag{3.15}$$

where $tg\delta = \varepsilon''/\varepsilon'$ is the loss tangent. The absorption coefficient is $q_e = 1/\ell$. Formula (3.15) is used for estimates of the penetration depth (or skin depth) of microwaves into a medium with certain complex permittivity.

The calculations by Equation 3.15 show that the depth of penetration of microwaves in the ocean water is equal to $\ell \approx (0.01-0.1)\lambda$, where λ is the wavelength in free space. In the range of wavelengths $\lambda = 0.3-3.0$ cm, value ℓ weakly depends on the temperature and salinity of the water. But in the range of $\lambda = 6.0-30$ cm, the depth of the skin layer depends essentially on salinity and temperature and can be equal up to several centimeters. Thus, the optimal range of wavelengths for remote measurements of sea surface

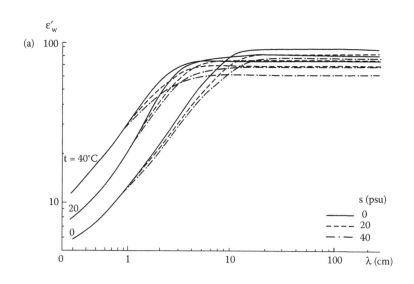

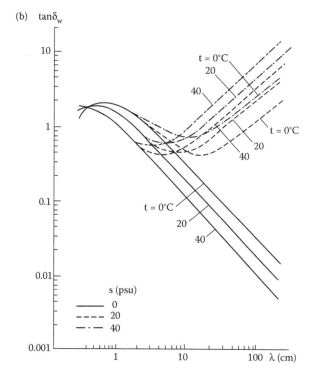

FIGURE 3.5
Complex permittivity of water versus electromagnetic wave. (a) Real part of complex permittivity. (b) Tangent of dielectric losses. Different values of temperature (t) and salinity (s) are denoted. (Cherny I. V. and Raizer V. Yu. *Passive Microwave Remote Sensing of Oceans*. 195 p. 1998. Copyright Wiley-VCH Verlag GmbH & Co. KGaA. Reproduced with permission.)

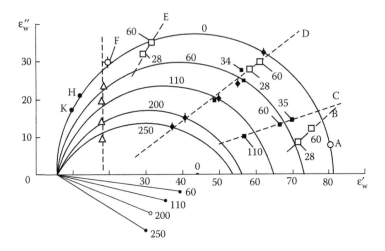

FIGURE 3.6
Cole–Cole diagrams for NaCl water solutions. Temperature: 20°C. Numbers are the values of salinity. Solid line—model calculations. Dashed line—measurements at the wavelengths: (A) λ = 17.24 cm; (B) λ = 9.22 cm; (C) λ = 10 cm; (D) λ = 3.2 cm; (E) λ = 1.26 cm; (F) λ = 0.8 cm; (H) λ = 0.5 cm; (K) λ = 0.4 cm. (Adapted from Sharkov E. A. 2003. *Passive Microwave Remote Sensing of the Earth: Physical Foundations.* Springer Praxis Books; Cherny I. V. and Raizer V. Yu. *Passive Microwave Remote Sensing of Oceans.* 195 p. 1998. Copyright Wiley-VCH Verlag GmbH & Co. KGaA. Reproduced with permission.)

temperature has to be λ = 3.0–8.0 cm, but for the measurements of sea surface salinity, it has to be λ = 18–75 cm.

At the same time, initial estimates based on the Fresnel equations and the Debye dielectric model have shown that the theoretical sensitivity (gradient) of the brightness temperature to minor variations of SSS and SST at C, S, and L bands is $(\partial T_B(t, s)/\partial s) \approx 0.2$–0.5 K/psu and $(\partial T_B(t, s)/\partial t) \approx 0.1$–0.2 K/°C, respectively. The sensitivity also depends on observation parameters, including microwave frequency, incidence angle, and polarization. Microwave remote sensing technology dedicated to the monitoring of SSS and SST is constantly advancing in order to obtain data with the highest accuracy (Wilson et al. 2001; Yueh et al. 2010, 2013; Yueh and Chaubell 2012).

3.3 Influence of Surface Waves and Wind

3.3.1 Introduction

The impact of the surface waves on the sea surface microwave emission has been studied by many researchers over the past 40 years. The first experiments have been made in the early 1970s (Hollinger 1970, 1971; Van Melle et al. 1973; Swift 1974). Microwave contributions from the surface waves were evaluated using first a one-scale and then a two-scale electromagnetic model (Wu and Fung 1972; Wentz 1975).

In particular, a two-scale model is designed as a superimposition of small- and large-scale surface irregularities affecting the scattering independently. Correspondingly, the methods of geometrical optics and the theory of small perturbations are invoked to compute microwave effects from large-scale and small-scale surface waves. These models operate with the Gaussian function of the wave slope distribution (or the Gram–Charlier series) with standard deviations dependent on the wind speed according to classical work (Cox and Munk 1954).

The application of the Gaussian law of wave slope distribution supposes that random surface irregularities can be presented by the statistical ensemble of linear flat waves. This is true when the long surface gravity waves on a deep water are considered in the low-frequency interval of spectral energy, that is, closely to spectral peak. Indeed, the contribution of such long-period gravity waves or swells into variations of microwave emission is small enough. The most abundant types of ocean surface waves are strongly nonlinear short gravity waves of finite amplitudes as well as highly nonlinear chaotic capillary waves of complex geometry. Their influence on microwave emission is more important for ocean diagnostics. The ensemble of nonlinear surface waves is not described only by the Gaussian law of distributions; therefore, several efficient models of scattering from random non-Gaussian surfaces have been considered and evaluated (Jakeman 1991; Tatarskii and Tatarskii 1996). Theoretically, the contributions from Gaussian and non-Gaussian wave statistics to the sea surface microwave emission can be distinguished using passive microwave radiometers (Irisov 2000).

The solution of practical problems of electromagnetic wave propagation across a rough statistical (random) surface is generally based on the asymptotic methods of diffraction theory and theory of electromagnetic wave propagation (Bass and Fuks 1979; Rytov et al. 1989; Ishimaru 1991; Voronovich 1999). More sophisticated technique supposes direct numerical solutions and simulations of Maxwell's equations, which can be applied in principle, for any surface geometry and/or statistical ensamples of surface waves. The corresponding codes and numerical examples of scattering and emission are reported in a book by Fung and Chen (2010). We believe that direct simulations of electromagnetic radiation fields could prove to be more valuable for remote sensing theoretical data than asymptotic solutions and/or approximations.

3.3.2 Resonance Theory of Microwave Emission of a Rough Water Surface

This theory was suggested and developed in order to investigate resonance effects (by analogy with Bragg resonance scattering) in thermal microwave emission from a small-scale sea surface. This theory is also known as "The Theory of Critical Phenomena in Microwave Emission of a Rough Surface" (Etkin et al. 1978; Kravtsov et al. 1978).

The very first approach was developed using the small-perturbation method allowing the authors to calculate the intensities of two diffraction maximums and the mirror reflective component of electromagnetic scattering from the one-dimensional and two-dimensional (cylindrical) sinusoidal dielectric surface. As a result, a simple analytical solution for the brightness temperature contrast due to small-scale periodic surface irregularities was obtained and tested experimentally (Irisov et al. 1987; Etkin et al. 1991; Trokhimovski et al. 2003; Sadovsky et al. 2009). Later, the model was updated (Yueh et al. 1994b; Irisov 1997, 2000; Johnson and Zang 1999; Johnson 2005, 2006; Demir and Johnson 2007) in order to obtain more precise solutions and explain better the impact of sea surface waves on the microwave emission.

According to the analytic theory within the limits of second-order perturbation theory, the brightness temperature contrast ΔT_B of a sinusoidal periodic surface (with respect to a smooth water surface) is defined as the following (Irisov 1987):

$$\Delta T_B \approx T_0 \cdot (k_0 a)^2 G\left(\frac{K}{k_0}, \varepsilon_w, \theta, \varphi, \tau_p\right), \tag{3.16}$$

where $k_0 = (2\pi/\lambda)$, $K = (2\pi/\Lambda)$; Λ and a are the wavelength and the amplitude of sinusoidal surface irregularities, respectively; $G(\ldots)$ is the resonance function dependent on the dielectric constant of the water $\varepsilon_w(\lambda)$, the angle of view from the nadir θ, the azimuth angle φ, and τ_p, which denotes polarization ($\tau_p = 0$ for vertical polarization and $\tau_p = \pi/2$ for horizontal polarization); and T_0 is the thermodynamic temperature of the water surface.

Because it is difficult to find in the literature a full set of analytical mathematical expressions for computing the resonance function $G(\ldots)$, we write these formulas (Irisov 1987; Raizer and Cherny 1994):

$$G = -\frac{1}{4}\text{Re}\left\{2(E^{(0)}E^{(2)*} + H^{(0)}H^{(2)*}) + \frac{c_+}{c_0}\left(\left|E_+^{(1)}\right|^2 + \left|H_+^{(1)}\right|^2\right) + \frac{c_-}{c_0}\left(\left|E_-^{(1)}\right|^2 + \left|H_-^{(1)}\right|^2\right)\right\}. \tag{3.17}$$

For zero-order scattered waves (specular reflected component)

$$E^{(0)} = U_0 E^{(i)} + W_0 H^{(i)}$$

$$H^{(0)} = V_0 H^{(i)} - W_0 E^{(i)}$$

$$V_0 = \frac{g_0 d_0 - (es_0)^2}{\omega_0}$$

$$U_0 = \frac{\tilde{g}_0 d_0 - (es_0)^2}{\omega_0}$$

$$W_0 = 2es_0c_0/\omega_0$$

$$g_k = c_k - f \cdot \tilde{c}_k$$

$$\tilde{g}_k = c_k - \varepsilon \cdot f \cdot \tilde{c}_k$$

$$d_k = c_k + f \cdot \tilde{c}_k$$

$$\tilde{d}_k = c_k + \varepsilon \cdot f \cdot \tilde{c}_k$$

$$\omega_k = d_k \tilde{d}_k + (es_k)^2$$

$$k = \begin{cases} + \\ 0 \\ - \end{cases}$$

For second-order scattered waves

$$E_\pm^{(1)} = \frac{d_\pm}{\omega_\pm} E_\pm^{(i)} + \frac{es_\pm}{\omega_\pm} H_\pm^{(i)}$$

$$H_\pm^{(1)} = \frac{d_\pm}{\omega_\pm} H_\pm^{(i)} - \frac{es_\pm}{\omega_\pm} E_\pm^{(i)}$$

where

$$E_\pm^{(i)} = E^{(0+i)}\left[\frac{\mu_0}{k_0^2}(\varepsilon - 1) + s_0 s_\pm (1 - \varepsilon \cdot f) - \varepsilon \cdot f \cdot \tilde{c}_0 \tilde{c}_\pm \right] - E^{0-i}\varepsilon \cdot f \cdot c_0 \cdot \tilde{c}_\pm + H^{(0+i)}es_\pm c_0$$

$$H_\pm^{(i)} = H^{(0+i)}\left[s_0 s_\pm (1 - f) - f \cdot \tilde{c}_0 \tilde{c}_\pm\right] - H^{(0-i)}f \cdot c_0 \cdot \tilde{c}_\pm - E^{(0+i)}es_\pm c_0$$

$$E^{(0\pm i)} = E^{(0)} \pm E^{(i)}$$

$$H^{(0\pm i)} = H^{(0)} \pm H^{(i)}$$

For second-order scattered waves (with a correction to the specular reflected component)

$$E^{(2)} = \frac{d_0}{\omega_0} E_2^{(i)} + \frac{es_0}{\omega_0} H_2^{(i)}$$

$$H^{(2)} = \frac{d_0}{\omega_0} H_2^{(i)} - \frac{es_0}{\omega_0} E_2^{(i)}$$

where

$$E_2^{(i)} = \varepsilon \cdot f \cdot \tilde{c}_0 \cdot (\tilde{c}_+ + \tilde{c}_- - 2\tilde{c}_0) \cdot [c_0 E^{(0-i)} + \tilde{c}_0 E^{(0+i)}]$$

$$+ \left[\frac{\mu_0}{k_0^2} - s_+ s_-\right] \cdot [c_0 E^{(0-i)} + \varepsilon \cdot f \cdot \tilde{c}_0 E^{(0+i)}]$$

$$- \varepsilon \cdot s_0 \cdot \left[H_+^{(1)} c_+ + H_-^{(1)} c_- + \left[\frac{\mu_0}{k_0^2} - s_+ s_-\right] \cdot H^{(0+i)}\right]$$

$$+ E_-^{(1)} \cdot \left[\varepsilon \cdot f \cdot \tilde{c}_0 (c_0 + \tilde{c}_0) - s_- s_0 (1 - \varepsilon \cdot f) - \frac{\mu_0}{k_0^2}(\varepsilon - 1)\right]$$

$$+ E_+^{(1)} \cdot \left[\varepsilon \cdot f \cdot \tilde{c}_0 (c_0 + \tilde{c}_0) - s_+ s_0 (1 - \varepsilon \cdot f) - \frac{\mu_0}{k_0^2}(\varepsilon - 1)\right],$$

$$H_2^{(i)} = f \cdot \tilde{c}_0 \cdot (\tilde{c}_+ + \tilde{c}_- - 2\tilde{c}_0) \cdot [c_0 H^{(0-i)} + \tilde{c}_0 H^{(0+i)}]$$

$$+ \left[\frac{\mu_0}{k_0^2} - s_+ s_-\right] \cdot [c_0 H^{(0-i)} + f \cdot \tilde{c}_0 H^{(0+i)}]$$

$$+ \left[\frac{\mu_0}{k_0^2} - s_+ s_-\right] \cdot [c_0 H^{(0-i)} + f \cdot \tilde{c}_0 H^{(0+i)}] + es_0 \left[E_+^{(1)} c_+ + E_-^{(1)} c_- + \left[\frac{\mu_0}{k_0^2} - s_+ s_-\right] \cdot E^{(0+i)}\right]$$

$$+ H_-^{(1)} [f \cdot \tilde{c}_0 (c_0 + \tilde{c}_0) - s_- s_0 (1 - f)] + H_+^{(1)} [f \cdot \tilde{c}_0 (c_0 + \tilde{c}_0) - s_+ s_- (1 - f)].$$

Coefficients are

$$\mu_0 = k_0^2 (1 - \sin^2\theta \sin^2\varphi)$$

$$\tilde{\mu}_0 = k_0^2 (\varepsilon - \sin^2\theta \sin^2\varphi)$$

$$s_\pm = \sin\theta \cos\varphi \pm k/k_0$$

$$s_0 = \sin\theta \cos\varphi$$

$$c_\pm = \left[\mu_0/k_0^2 - s_\pm^2\right]^{1/2}$$

$$\tilde{c}_\pm = \left[\tilde{\mu}_0/k_0^2 - s_\pm^2\right]^{1/2}$$

$$c_0 = \cos\theta$$

$$f = \mu_0/\tilde{\mu}_0$$

$$e = (1-f)\sin\theta\sin\varphi.$$

Incident electric and magnetic fields are introduced in normalized form:

$$E^{(i)} = \sin(\chi_0 + \tau_s)$$

$$H^{(i)} = \cos(\chi_0 + \tau_s)$$

$$tg\chi_0 = tg\varphi\cos\theta.$$

With the constraint

$$2n\frac{K}{k_0}\sin\theta\cdot\cos\varphi + \left(n\frac{K}{k_0}\right)^2 = \cos^2\theta, \tag{3.18}$$

resonance effects in the microwave emission from sinusoidal surface appear. It is important to note that resonance conditions (3.18) for thermal microwave emission differ from the Bragg resonance conditions for scattering. As follows from Equation 3.18, resonance maximums of the first order $n = \pm 1$ are realized better at the nadir angle of view. This model provides the possibility of separating the influence of different polarization on the microwave emission at the nadir, and at the grazing angles of view.

Figure 3.7 shows examples of calculated resonance function G(...) for different parameters of the model. Numerical results show that the value of the brightness temperature contrast due to the influence of small-scale surface sinusoidal irregularities can be reached approximately to $\Delta T_B \approx 30$ K at the resonance maximums. The effects were studies using different microwave radiometers in the laboratory and an agreement between the theory and experiment was shown (Trokhimovski et al. 2003).

Within the limits of the approach of small perturbations, the theory can be modified for a statistical surface with 2-D wave number spectrum of roughness. In this case, relationship (3.16) takes the form:

$$\Delta T_B = 2T_0 k_0^2 \int_0^{2\pi}\int_{K_{min}}^{\infty} G\left(\frac{K}{k_0}, \varepsilon_w, \theta, \varphi\right)\cdot S(K,\varphi)K dK d\varphi, \tag{3.19}$$

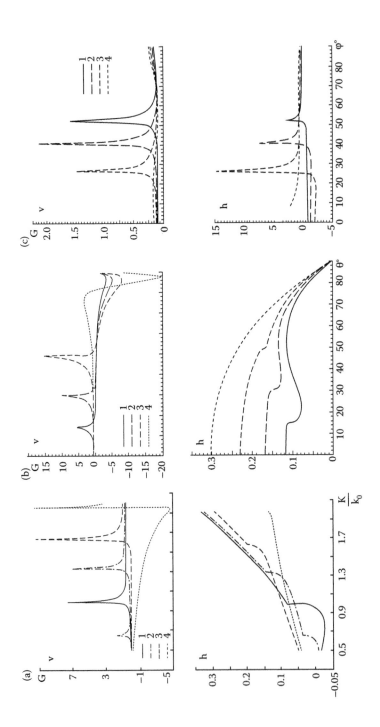

FIGURE 3.7

Critical phenomena in thermal microwave emission of a rough surface. Resonance functions $G(K/k_0; \theta, \varphi)$ computed at the wavelength $\lambda = 2$ cm for vertical (v) and horizontal (h) polarizations. (a) $G(K/k_0)$; $\varphi = 0°$. View angle is changed: (1) $\theta = 0°$; (2) $\theta = 20°$; (3) $\theta = 40°$; (4) $\theta = 70°$. (b) $G(\theta)$; $\varphi = 0°$. (1) $K/k_0 = 1.25$; (2) $K/k_0 = 1.5$; (3) $K/k_0 = 1.75$; (4) $K/k_0 = 2.0$. (c) $G(\varphi)$; $\theta = 70°$. (1) $K/k_0 = 1.25$; (2) $K/k_0 = 1.5$; (3) $K/k_0 = 1.75$; (4) $K/k_0 = 2.0$. (From Cherny I. V. and Raizer V. Yu. *Passive Microwave Remote Sensing of Oceans*. 195 p. 1998. Copyright Wiley-VCH Verlag GmbH & Co. KGaA. Reproduced with permission.)

and $S(K, \varphi)$ is the directional wave number spectrum of a rough surface

$$S(K,\varphi) = \frac{1}{K}F(K)Q(\varphi - \varphi_0)$$

$$Q(\varphi - \varphi_0) = \frac{1}{2\pi}[a + b(K)\cos[2(\varphi - \varphi_0)]],$$

where $F(K)$ is the omnidirectional wavenumber spectrum (Chapter 2), $Q(\varphi - \varphi_0)$ is a nondimensional spreading function, φ_0 is the azimuth angle of observations, K_{min} is the low-frequency cutoff (usually $K_{min} = 0.05k_0$), and coefficients are $a = 1$ and $b(K) \approx 0.5$. According to expansion (Irisov 1997, 2000), resonance model (3.19) describes the microwave emission effects from both small-scale and large-scale surface waves. For example, a set of parameterizations (2.1) through (2.10) of wave number spectrum $F(K)$ (Chapter 2) can be employed to compute the brightness temperature contrast $\Delta T_B(V)$ dependent on wind speed V.

To understand the behavior of brightness contrast ΔT_B, the power wave number spectrum $S(K, \varphi) = AK^{-n}Q(\varphi)$, where A and n are parameters, is incorporated in Equation 3.19. Calculations are performed in a wide range of parameters $A = 10^{-4}-10^{-2}$ and power exponent $n = 2-5$. Calculations (Figure 3.8) show that the integration (3.19) gives certain smoothing to the resonance maxima. Moreover, the change of the power exponent n in the spectrum results in

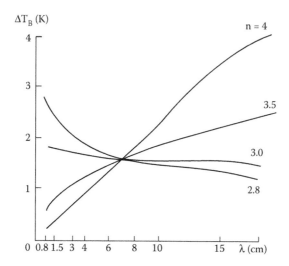

FIGURE 3.8

Brightness temperature contrast of a rough sea surface (at nadir view angle) computed using resonance model (3.19) with wave number spectrum $F(K, \varphi) = AK^{-n}Q(\varphi)$. Power exponent is varied: $n = 2.8$; 3.0; 3.5; 4 (denoted). $A = 10^{-4}$. (Cherny I. V. and Raizer V. Yu. *Passive Microwave Remote Sensing of Oceans*. 195 p. 1998. Copyright Wiley-VCH Verlag GmbH & Co. KGaA. Reproduced with permission.)

considerable variations of spectral dependency $\Delta T_B(\lambda)$ that is explained by the selective character of the contributions from different surface harmonics.

The resonance model (3.19) has some modifications providing a good agreement with field experimental radiometric data obtained from the sea platform and aircraft laboratory. The model also creates a physical basis for *radio-spectroscopy* of gravity–capillary surface waves (Irisov et al. 1987; Etkin et al. 1991; Irisov 1991; Trokhimovski 2000). Finally, the model explains the effect of polarization anisotropy in the ocean microwave emission manifested first in work (Dzura et al. 1992) and investigated later in more detail (Yueh et al. 1994a, 1995, 1997, 1999; Pospelov 1996, 2004; Skou and Laursen 1998; Trokhimovski et al. 2000; Laursen and Skou 2001). Today, the principle of polarization anisotropy (or polarimetric microwave radiometry) is used successfully for airspace measurements of ocean surface wind vector.

3.3.3 Two-Scale and Three-Scale Modified Models

Further development of the microwave emission theory is connected with the modification of a two-scale model. The improvement accounts for the resonance character of the microwave contributions from small-scale wave components according to Equation 3.19. The brightness temperature contrast is now written as

$$\Delta \bar{T}_B = \int\limits_{-\infty}^{+\infty} \int\limits_{-\infty}^{+\infty} \Delta T_B(z_x, z_y) P_\theta(z_x, z_y) dz_x dz_y, \tag{3.20}$$

where $P_\theta(z_x, z_y)$ is the probability distribution function of wave slopes (in local coordinate system) which can be represented as the Gaussian or non-Gaussian law.

It was shown theoretically (Voronovich 1994, 1996) that the diffraction of the incident electromagnetic field on large-scale components of surface irregularities cannot be considered using the Kirchhoff approximation at low grazing ($\theta > 70°$ from nadir) view angles. In this case, the curvature of flat facets becomes an important regulating parameter in a two-scale surface model.

At the same time, the comparison of resonance theory and Kirchhoff approximation for microwave emission shows that both methods yield identical results (Irisov 1994, 1997). It means that the resonance model (3.19) can describe the contributions from both large- and small-scale surface irregularities having small slopes. For correct computations of the contrast ΔT_B, it is enough to shift the value of cutoff K_{min} to a more low-frequency range (to set $K_{min} = 0.05 \, k_0$); it will correspond to an ensemble of large gravity waves with small slopes. However, at the grazing view angles, one needs to take high-order terms in the small-slope approximation into account to obtain the correct results.

The next step in applying the microwave emission theory includes the creation of a three-scale model. In this model, the following independent parts are involved: the statistical ensemble of large-scale gravity waves (by geometric optics) + gravity–capillary waves (by resonance model) + small-scale nonlinear waves with a large steepness (by quasi-static model). Such a model can describe the impact of multiscale surface waves on the microwave emission at C, S, and L bands ($\lambda = 8$–30 cm) where the penetration depth is larger than at K band. Moreover, it is possible to invoke a quasi-static (imped-ance) approach to describe the macroscopic properties of a rough air–sea interface (Section 3.3.5). In this case, we obtain a three-scale model

$$\Delta \bar{T}_B = \int\limits_{-\infty}^{+\infty}\int\limits_{-\infty}^{+\infty} \Delta T_{Bres}(\varepsilon_{eff}; z_x, z_y) P_\theta(z_x, z_y) dz_x dz_y. \tag{3.21}$$

in which contrast ΔT_{Bres} is calculated by resonance model (3.19) and the per-mittivity of the water is replaced by the effective permittivity, that is, $\varepsilon_w \to \varepsilon_{eff}$. In the case of steep irregularities, the effective permittivity of the air–sea inter-face ε_{eff} is calculated through moments of the surface wave number spectrum (Kuz'min and Raizer 1991; Cherny and Raizer 1998). Although such a three-scale model requires more detailed investigations, we believe that it will be useful for the interpretation of *complex multiscale hydrodynamic signatures* asso-ciated with roughness change in the field of strong (sub)surface turbulence.

3.3.4 Contribution of Short Gravity Waves

The microwave effects from a statistical ensemble of nonlinear short gravity waves on deep water (known as the *finite amplitude waves*) can be investi-gated using geometric optics approximation. The key parameter here is a non-Gaussian probability density function (PDF) of wave slopes. To illus-trate this statement, let us consider a one-dimensional multimode random process describing a rough surface in the form:

$$\xi(x) = \sum_{n=1}^{N} a_n \cos(K_n x + \psi_n), \tag{3.22}$$

where a_n and K_n are the amplitude constants and spatial frequencies of the harmonics (wave modes), ψ_n are the random phases, which are uniformly distributed over the interval $[0, 2\pi]$, and n is the number of surface harmon-ics. To determine the surface emissivity, it is necessary to define PDF of the derivatives (surface slopes) of this process. In the case of nonsynchronized phases, the PDF is

$$P(z) = \frac{1}{2\pi} \int\limits_{-\infty}^{+\infty} \prod_{n=1}^{N} J_0(Uz_n) e^{-iUz} dU, \tag{3.23}$$

where $J_0(x)$ is the zeroth-order Bessel function of the real argument, $z_n = K_n a_n = nC_n(Ka)^n$, and Ka is the initial wave steepness. Formula 3.23 is obtained using the characteristic function of multimode random process (Akhmanov et al. 1981). The nonlinearity is introduced via the coefficients a_n of the Stokes expansion. For harmonics $N = 1, 2, 3, 4$, the distribution differs significantly from the Gaussian distribution, and begins to approach it only for $N \geq 5$ (Figure 3.9). The emissivity is now found by the averaging procedure:

$$\kappa(\theta) = 1 - \int_{-\infty}^{+\infty} |r(\cos\chi)|^2 P(\theta, z) dz, \qquad (3.24)$$

where $r(\cos\chi)$ is the Fresnel reflection coefficient for the vertical or horizontal polarization; $\chi = \chi(z, \theta)$ is a local incident angle, which depends on the angle of view θ and the slope of the surface wave z. As a result, the brightness temperature contrast (relative to a smooth surface) is a function of wave slope and number of harmonics $\Delta T_B(z, N)$, that is, depends on nonlinearity of the surface

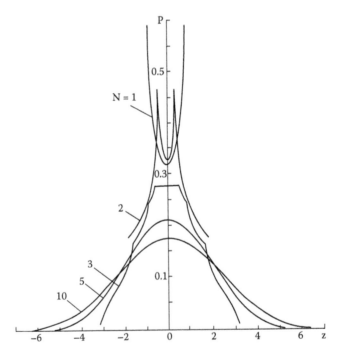

FIGURE 3.9
Transformation of wave slope's PDF for multimode non-Gaussian random surface. Number of modes is varied: $N = 1; 2; 3; 5; 10$. Initial wave steepness $Ka = 0.75$. (Cherny I. V. and Raizer V. Yu. *Passive Microwave Remote Sensing of Oceans*. 195 p. 1998. Copyright Wiley-VCH Verlag GmbH & Co. KGaA. Reproduced with permission.)

wave. Formulas (3.24) and (3.25) yield estimates of the contribution from non-Gaussian wave slope statistics to the sea surface microwave emission.

To verify this model, unique remote sensing studies of short nonlinear gravity waves were conducted in outdoor water tank using highly sensitive passive microwave radiometers (created on the superconducting Josephson detector) at the wavelengths $\lambda = 0.8$ and 1.5 cm (Il'in et al. 1985, 1988, 1991; Ilyin and Raizer 1992). These measurements have shown linear dependencies of the brightness temperature contrast ΔT_B on the amplitude of the surface waves and demonstrated a good agreement between model and experimental data.

Moreover, the results reveal the strong dependence of the brightness temperature contrast on the steepness of surface waves. Both theoretical and experimental data demonstrate a high sensitivity of microwave emission to the geometry and nonlinearity of short gravity waves (Figure 3.10a and b). The minimal value of the brightness contrast from weakly nonlinear surface waves is about $\Delta T_B = 0.2$ K. When the steepness of surface waves increased, the brightness temperature contrast reached the value of $\Delta T_B = 8{-}10$ K at both wavelengths of emission $\lambda = 0.8$ and 1.5 cm.

Similar calculations of the microwave emission can be made using one-scale model and the Gaussian slope approximation (Cox and Munk 1954). In

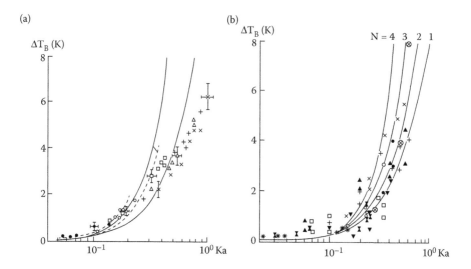

FIGURE 3.10

Brightness temperature contrast due to short nonlinear surface gravity waves. Dependencies ΔT_B(Ka) at wavelength $\lambda = 1.5$ cm and incidence angle $\theta = 30°$, vertical polarization. Experimental data (symbols) are shown for different surface wavelengths. Calculations: (a) one-mode $N = 1$ and two-mode $N = 2$ approximations. Solid line—data without atmospheric contribution; dashed line—data with atmospheric contribution. (b) $N = 1$; 2; 3; 4. (Adapted from Il'in V. A. et al. 1985. *Izvestiya, Atmospheric and Oceanic Physics.* 21(1):59–63 (translated from Russian); Il'in V. A. et al. 1988. *Izvestiya, Atmospheric and Oceanic Physics.* 24(6):467–471 (translated from Russian); Ilyin V. A. and Raizer V. Yu. 1992; Cherny I. V. and Raizer V. Yu. *Passive Microwave Remote Sensing of Oceans.* 195 p. 1998. Copyright Wiley-VCH Verlag GmbH & Co. KGaA. Reproduced with permission.)

this case, the mean value of large-scale gravity wave's slopes is Ka ~ 10^{-3}–10^{-2}. However, in the case of finite-amplitude surface waves, this value may be Ka ~ 0.5–1.0, and microwave emission effect due to a wave's nonlinearity seems to be more important than that due to Gaussian slope distribution. Moreover, experimental observations have demonstrated the existence of steady three-dimensional symmetric water wave patterns, which are the result of bifurcation of the Stokes waves of large steepness Ka ≥ 0.25 (Chapter 2). Therefore, more reliable analysis of microwave emission at pre-bifurcation conditions should be made using two- or three-dimensional multimode surface model, for example, written in the following form:

$$\xi(x,y) = \sum_{n}\sum_{m} A_{n,m}(K_x, K_y)\cos(nK_x x + \varphi_n)\cos(mK_y y), \qquad (3.25)$$

where $A_{n,m}$ are the amplitudes of the modes, and K_x and K_y are the surface wave numbers. The numbers of harmonics in sum (3.25) can be varied in order to reveal microwave effects associated with the nonlinearity of surface waves and non-Gaussian statistics of wave slopes. Brightness temperature contrast $\Delta T_B = (\kappa - \kappa_0)T_0$ from an ensemble of nonlinear surface waves (3.25) can be calculated using formula (3.24) as a function $\Delta T_B[P(z)]$, where $P(z)$ is the PDF of derivatives $z = (\partial\xi/\partial x, \partial\xi/\partial y)$. The PDF is generated numerically.

3.3.5 Quasi-Static and Impedance Models

Under certain conditions ($k_0 \ll K$, $k_0 a \ll 1$, where $K = 2\pi/\Lambda$, $k_0 = 2\pi/\lambda$, Λ and a are the horizontal and vertical dimensions of the irregularities), a random or deterministic air–sea interface can be represented by transition dielectric structures with effective parameters. An exact analytic solution of the impedance electromagnetic problem that is especially adapted to stochastic and multiscale rough sea surface is complicated as follows from the book Bass and Fuks (1979); it might not always be easy to use for the analysis and interpretation of microwave data. Therefore, in order to compute sea surface emissivity, a numerical method, based on the electromagnetic theory of layered media (Stratton 1941; Brekhovskikh 1980), can be applied. Multilayer dielectric models and algorithms were used in remote sensing studies, (for example, Raizer and Cherny 1994; Cherny and Raizer 1998; Sharkov 2003; Franceschetti et al. 2008; Imperatore et al. 2009; Lin et al. 2009) and other environmental tasks.

As a whole, a quasi-static macroscopic model provides the so-called impedance matching mechanism between air and water media that yields significant increases in microwave emission. The impact depends on parameters and configuration of the air–sea interface.

A quasi-static effect from surface roughness has been tested first in the laboratory using a microwave radiometer at the wavelength $\lambda = 18$ cm (Gershenzon et al. 1982). Surface irregularities on the water surface were created in the laboratory tank with the aid of a foam radiotranslucent sheet with sinusoidal (or rectangular) profiles of different amplitudes and spatial periods. The corrugated side of the sheet was pushed into the water with the smooth side turned to the antenna of the radiometer. In this manner, the "frozen" regular structure with the various and well-known parameters and geometry is reproduced. Comparison between test experimental data and model calculations shows their good agreement.

The application of the macroscopic theory for ocean remote sensing is a little bit complicated because of the complexity of the interface profiles generated by 2-D and 3-D small-scale surface waves. Theoretically, macroscopic models of a rough random surface are justified by the inadequacy of perturbation's methods for the analysis of steep and closely spaced surface irregularities. The macroscopic model places no constraints on the parameter of wave steepness Ka > 1. This allows for a description of microwave emission of very unstable capillary waves (steep ripples) having impulse-type configurations. The model yields the higher brightness temperature contrasts depending on the structure and effective parameters of the transition layer that is important for the detection of surface roughness anomalies at S–L bands (Raizer 2014).

The comparison of two electromagnetic microwave emission models—"resonance" and "macroscopic"—for the same sinusoidal surface $z = a\cos(Kx)$, reveals the limits of their applicability by the parameter of wave steepness Ka. Three different methods were used for the calculation: (1) the resonance model, (2) quasi-static macroscopic model, and (3) numerical solution of the diffraction problem based on a method of integral equations (Petit 1980).

It was found (Cherny and Raizer (1998) that there is a connection between resonance and macroscopic models at the wave steepness parameter Ka ~ 0.5 for dependencies $\Delta T_B(Ka)$ calculated at the wavelength $\lambda = 18$ cm.

In the case of resonance model, contrast ΔT_B is determined solely by the value of the parameter $k_0 a$, and is practically independent of the steepness Ka. Note that in the region Ka > 1, the method of small perturbations is unsuitable and a numerical solution of the diffraction problem can give another maxima in the microwave emission.

In the case of the macroscopic model ($k_0 \ll K$, $k_0 a \ll 1$), the brightness temperature contrast tends asymptotically to zero $\Delta T_B \to 0$ as Ka $\to \infty$ (actually for Ka > 100), which corresponds to the case of closely packed and very steep small-scale surface irregularities.

In view of macroscopic theory and remote sensing, surface roughness and the corresponding model transition structure should have equivalent electromagnetic responses. It can be reached when the radius of surface curvature $R \sim 1/(K^2 a)$ and the thickness of electromagnetic skin layer $L \sim 1/(k_0\sqrt{|\varepsilon_w|})$ are compatible with each other, that is, $R \sim L$. This relationship leads to the following general conditions:

$$(Ka)^2 \ll \sqrt{|\varepsilon_w|}, \quad k_0 a \ll 1, \quad k_0 \ll K, \qquad (3.26)$$

which are fully satisfied at S–L band, (for example, for electromagnetic wavelengths $\lambda \geq 6$–8 cm, $|\varepsilon_w| \approx 30$–80) and any small-scale sea surface waves with steepness/roughness parameter $(Ka) \geq 1$ (where $k_0 = 2\pi/\lambda$ and $K = 2\pi/\Lambda$ are electromagnetic and surface wave numbers, respectively; a and Λ are vertical and horizontal scales of roughness; and ε_w is the dielectric constant of seawater).

In some cases, the microwave properties of a steep rough air–water interface can be modeled by a multilayer dielectric structure with the complex characteristic impedance, which is defined as follows (Stratton 1941; Brekhovskikh 1980):

$$z_i = \eta_i \frac{z_{i+1} + j\eta_i \tan k_i h_i}{\eta_i + j z_{i+1} \tan k_i h_i}, \quad i = 0, 1, 2, \ldots, N, \tag{3.27}$$

$$\eta_i = \begin{cases} \eta_0 / \left(\sqrt{\varepsilon_i}\, \cos\theta_i\right) & \text{for perpendicular (vertical) polarization} \\ \eta_0 \cos\theta_i / \sqrt{\varepsilon_i} & \text{for parallel (horizontal) polarization} \end{cases}$$

is the intrinsic impedance of the i-th layer; $\eta_0 = 120\pi$ ohms is the wave impedance in free space; $\varepsilon_i = \varepsilon_i' + j\varepsilon_i''$ is the complex dielectric constant for the i-th layer; $k_i = (2\pi/\lambda)\sqrt{\varepsilon_i}\, \cos\theta_i$ is the wave propagation constant; h_i is the thickness of the i-th layer; θ_i is the incident angle for the i-th layer; N is the total number of layers; and λ is the electromagnetic wavelength.

Spectral reflection and emission coefficients of a multilayer dielectric structure (for vertical and horizontal polarizations) are defined through the characteristic impedances of layers as follows:

$$r_{i\lambda} = \frac{z_{i+1} - z_i}{z_{i+1} + z_i}, \quad \kappa_{i\lambda} = 1 - |r_{i\lambda}|^2, \quad i = 0, 1, 2, \ldots, N. \tag{3.28}$$

The resulting complex impedance z_N is computed using the layer recursions. Basically, it depends on the number of input layers N involved in a multilayer structure. Complex reflection coefficient $r_{N\lambda}$ and emission coefficient $\kappa_{N\lambda}$ (emissivity) are also computed for the same number N.

Let us consider several variants of quasi-static microwave models.

3.3.5.1 A Single Dielectric Slab

It is the simplest variant of the impedance model that is widely used in microwave remote sensing. In this case, a random rough air–water interface is described by effective complex permittivity

$$\varepsilon_{eff} \approx (1 - \bar{c})\varepsilon_a + \bar{c}\varepsilon_w, \quad \bar{c} = 1 - 2\int_0^h \Phi(z/\sigma_\xi)dz, \quad 0 < \bar{c} < 1, \tag{3.29}$$

where ε_a and ε_w are the dielectric constants of air and water, respectively, $\bar{c}$ is the filling coefficient (mean bulk concentration of water in the slab), $\Phi(x)$ is the probability integral, and z is the vertical coordinate. The filling coefficient $\bar{c}$ is computed using the double integral as the volume under generated surface $z_\xi = \xi(x, y)$ with variance $\sigma_\xi^2 = (1/2\pi)^2 \int F(\bar{K})d\bar{K}$, where $F(\bar{K})$ is the directional wave number spectrum of the sea surface. The variance yields the "effective" thickness of the interface $h \approx \sqrt{\sigma_\xi^2}$, which can be related to the total thickness of a multilayer structure (Figure 3.11a,b).

In the case of a Gaussian random isotropic surface, the filling coefficient can be defined using the theory of excursions of a random field (Bunkin and Gochelashvili 1968; Belyaev and Nosko 1969; Nosco 1980). Spectral emissivity is computed using the complex Fresnel reflection coefficients, derived for a uniform dielectric slab (Landau and Lifshitz 1984; Born and Wolf 1999) with input parameters $\{h_1 = h, \varepsilon_1 = \varepsilon_{eff}\}$. As a whole, the impedance model

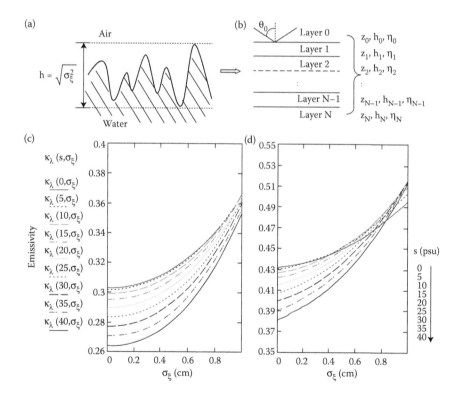

FIGURE 3.11
Contribution of surface roughness to microwave emission at L band ($\lambda = 21$ cm). (a) Illustration of surface impedance model and (b) multilayer approach. Emissivity $\kappa_\lambda(s,\sigma_\xi)$ computed as function of r.m.s. of the surface elevation σ_ξ for (c) horizontal and (d) vertical polarizations. Salinity is varied: $s = 0, 5, 10, 15, 20, 25, 30, 35$, and 40 psu (marked). Incidence 37°. Surface temperature $t = 10$°C.

describes the averaged microwave emission effects due to random small-scale surface roughness depending on the variance σ_ξ^2.

Examples of calculated emissivity at $\lambda = 21$ cm for different values of sea surface salinity using the impedance model are shown in Figure 3.11c and d (Raizer 2014). From these data, it follows that the impact of small-scale random surface roughness with steep-slope irregularities (for example, gravity–capillary and capillary waves) can be compatible with salinity variations, at least, in the interval of $s = 20$–30 psu. However, in practice, microwave effects induced by surface roughness, that is, due to variations of σ_ξ, can be eliminated somehow as a constant trend of the brightness temperature. Therefore, the impedance-based approach is very useful for providing a roughness-change correction needed for a better retrieval of SSS by microwave data.

3.3.5.2 Matching Transition Layer

This is the other modification of the impedance model. This variant represents a smooth transition layer with the following dielectric profile (Epstein's transition layer):

$$\varepsilon(z) = \frac{1}{2}(\varepsilon_a + \varepsilon_w) + \frac{1}{2}(\varepsilon_a - \varepsilon_w)\tanh\left(\frac{mz}{2}\right), \quad 0 < z < h. \tag{3.30}$$

The transition layer (3.30) provides a perfect broadband impedance matching between two dielectric media (air and water) with strongly nonuniform interface. As a result, considerable variations of emissivity at the S–L band occur. The important characteristics here are the total thickness (h) and the matching coefficient (m), which can be functions of surface parameters; they can be parameterized by wind speed as well. Emissivity is computed numerically using a method of layer recursions operated with discrete complex vertical profile $\varepsilon(z_i)$. The transition model can also be useful for the estimation of microwave emission effects induced by mixed roughness-volume irregularities occurring, for example, at high winds (Chapter 2).

3.3.5.3 Stochastic Multilayer Structural Model

This is an electromagnetic random field macro-model describing the microwave properties of the mixed air–sea interface of complex geometry. A stochastic model provides a multiple matching between many hydrodynamic factors and surface nonuniformities. Actually, it is an ideal phenomenological conceptual approach for supporting global remote sensing observations of the oceans using low-resolution S–L band imagery. This model does not require invoking any real-time geophysical information or additional data that may not always be available. Numerical realizations of the model are based on the generation of a large number of input physical parameters having certain statistical distributions and relationships. The generation process can be organized using Monte Carlo methods or other statistical network algorithms. As a result, it might be possible to reveal and estimate the appearance of multicontrast *stochastic microwave signatures* (especially at S–L bands) associated with different hydrodynamic phenomena or events.

Macroscopic models and modifications can be used in ocean remote sensing as an alternative and efficient approach to perturbation-based wave propagation theory as it relates to scattering and emission of low-frequency radiowaves from a rough sea surface. In particular, impedance-based models can explain measurable low-contrast and short-term variations (fluctuations) of the brightness temperature $\Delta T_B \sim 1$–2 K induced by steep capillary and strongly nonlinear gravity–capillary waves having an irregular profile (Cherny and Raizer 1998; Raizer 2014). Moreover, the impedance approach may have some advantages in global methods of surface roughness correction at the retrieval of sea surface salinity using L band space-based observations.

3.3.6 Influence of Wind

Wind speed as a geophysical parameter is often used in remote sensing in order to characterize the observed variations of the sea surface brightness temperature. Dependencies of the brightness temperature on wind speed or the so-called spectral microwave radiation-wind dependencies have been measured and investigated by many authors during the years (Shutko 1986; Sasaki et al. 1987; Goodberlet et al. 1989; Hollinger et al. 1990; Liu et al. 1992, 2011; Wentz 1992; 1997; Liu and Weng 2003; Bettenhausen et al. 2006; Yueh et al. 2006, 2013; Meissner and Wentz 2012). These data are well known; they are used in many applications involving the wind vector retrieval algorithms.

As an example, we refer to aircraft microwave radiometric observations conducted in the Pacific Ocean in 1986–1991 (Irisov et al. 1987; Etkin et al. 1991; Trokhimovski et al. 1995). In particular, during multifrequency measurements, correlations between different radiometric data were manifested at variable wind conditions. The most interesting results were obtained using two-channel radiometer system operated at wavelengths $\lambda = 8$ and 18 cm with a fluctuation sensitivity 0.1 and 0.15 K, respectively (Bolotnikova et al. 1994).

Figure 3.12 shows experimental two-channel regression between brightness temperature contrasts ΔT_{B18} ($\lambda = 18$ cm) and ΔT_{B8} ($\lambda = 8$ cm) plotted for three averaged sea state gradations. These data show that the regression coefficient $\rho = \Delta T_{B8}/\Delta T_{B18}$ changes considerably depending on sea surface state. The calculated values are $\rho = 2.30$, 1.76, and 1.22 for the 1–2, 3–4, and 5–6 of the Beaufort force, respectively. Because atmospheric effects at these wavelengths can be neglected, the observed variations of the brightness temperature are associated with the roughness change. The main dynamical factor here is wind-dependent wave spectrum and therefore, the regression is parameterized by wind speed. Analysis of airborne radiometric data show

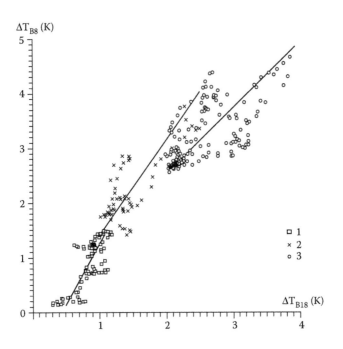

FIGURE 3.12

Experimental two-channel regression between brightness temperature contrasts ΔT_{B18} ($\lambda = 18$ cm) and ΔT_{B8} ($\lambda = 8$ cm) plotted for three averaged sea state gradations: "1"—1...2; "2"—3...4; "3"—5...6 of the Beaufort wind force. Antonov An-30. Pacific Ocean. (Cherny I. V. and Raizer V. Yu. *Passive Microwave Remote Sensing of Oceans*. 195 p. 1998. Copyright Wiley-VCH Verlag GmbH & Co. KGaA. Reproduced with permission.)

that the macroscopic model can give a good agreement between theoretical and experimental dependencies $\Delta T_B(V)$ at wavelengths $\lambda = 8$ and 18 cm only.

On the other hand, variations in the regression coefficient between data obtained from two radiometric channels $\lambda = 8$ and 18 cm, can be easily estimated theoretically. For this wave number spectrum, $F(K) = AK^{-n}$ is incorporated into the resonance model (3.19). In the linear approximation and at nadir view angle $\theta = 0$, the following expression for the regression coefficient is obtained:

$$\rho(n) = \left[\frac{k_{01}}{k_{02}}\right]^{-n+3}, \quad n \neq 1. \tag{3.31}$$

This relationship between the regression coefficient and the exponent of the power-type spectrum is simplest because resonance functions $G(K/k_{01})$ and $G(K/k_{02})$ in formula (3.16) are equal to each other practically at the wavelengths $\lambda_1 = 8$ cm and $\lambda_2 = 18$ cm (but wave numbers are different: $k_{01} = 2\pi/\lambda_1 = 0.785$ cm^{-1}; $k_{02} = 2\pi/\lambda_2 = 0.350$ cm^{-1}). The slope angle of the regression is easy to determine from Equation 3.31 and it is equal to

$\psi = \tan^{-1}[\rho(n)]$. In this particular case, the calculated regression coefficient $\rho = \Delta T_{B8}/\Delta T_{B18}$ fits experimental data the best at values of the power exponent equal to $n = 2$–3.

The detailed theoretical analysis of the presented experimental data has shown that both resonance narrowband mechanism and macroscopic broadband mechanism can contribute to ocean microwave emission. The first mechanism provides predominantly in the initial stage of wind-wave generation, when the different regular (periodic) wave structures are formed on the ocean surface at $V < 5$ m/s. The second mechanism is more efficient at higher winds $V > 7$–10 m/s when wave breaking processes are started and the wave's structures become more chaotic and unpredictable.

Actually, the choice of the model depends on the purpose of the remote sensing experiment, and knowledge of the ocean and atmosphere conditions. Thus, for more accurate theoretical analysis of multiband radiation-wind dependencies, the application of the resonance models is preferable.

This statement has been clearly illustrated long time ago in terms of the radiation-wind sensitivity $\Delta T_B/\Delta V$. Figure 3.13 shows simulated numerically dependencies $\Delta T_B/\Delta V$ versus λ using a combined ocean–atmosphere

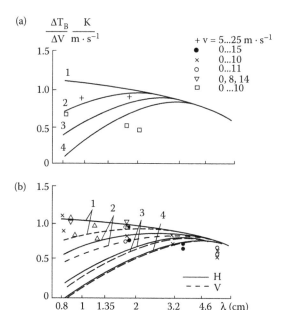

FIGURE 3.13
Spectral dependencies of the radiation-wind sensitivity in the ocean–atmosphere system computed using combined numerical algorithm. View angle: (a) $\theta = 0°$; (b) $\theta = 50°$. "H" and "V" denote horizontal and vertical polarization. Atmosphere parameters: water vapor is varied: (1) 0; (2) 0.71; (3) 2.0; (4) 3.5 kg/m³; humidity is 14.9 kg/m³ (constant). Symbols denote experimental data. (Cherny I. V. and Raizer V. Yu. *Passive Microwave Remote Sensing of Oceans.* 195 p. 1998. Copyright Wiley-VCH Verlag GmbH & Co. KGaA. Reproduced with permission.)

radiation model (Kosolapov and Raizer 1991). The sensitivity is defined not only by a large number of input geophysical parameters but also depends on the specifications of the sea surface microwave model.

In the mid-2000s, a new level in remote sensing of the ocean has been achieved due to innovations in space-based microwave technology. First of all, missions such as ESA SMOS (Kerr et al. 2001) and NASA Aquarius (Le Vine et al. 2007) dedicated to the global monitoring of SSS and SST have provided significant progress in microwave radiometry, especially at S and L band. Although a review of the available literature of SMOS and Aquarius data is beyond the scope of this book, however, some test radiometric experiments conducted are important for physics-based analysis and validation of microwave emission models.

Among them, we emphasize upon the field radiometric experiments supported by the ESA SMOS (Camps et al. 2002, 2003, 2004, 2005; Etcheto et al. 2004). In these experiments, accurate measurements of radiation-wind dependencies were made using L band radiometer. The data obtained show that there can be an uncertainty in the SSS retrieval due to the variability of sea surface conditions.

Figure 3.14 demonstrates the performance of these L band observations. To provide an accurate analysis, we compare experimental (Camps et al. 2002) and model (Raizer 2001, 2009) data. The radiation-wind dependencies are computed for different values of SSS in order to fit experimental data. The range of parameters SSS and SST corresponds to local sea conditions and *in situ* measurements. As a whole, there is good agreement between model and experimental microwave data. Radiation-wind dependencies are well distinguished by SSS; however, it seems that the contributions from wind action and salinity (and temperature) can be comparable to each other in some cases. It is typical for high-resolution (~0.3–0.5 km) observation situation when fluctuations in radiometric signal were registered at increasing wind and in the presence of wave breaking events. Indeed, fluctuations and trends in microwave radiometric signals are defined mostly by surface dynamics (but not by salinity or temperature). The roughness change and breaking waves yield considerable contributions into the sea surface microwave emission at L band. The performance of bias correction methods can be improved significantly by using multiband radiometric measurements that are sensitive to different environmental factors. This option may also reduce errors at the retrieval of SSS (and SST).

3.4 Impact of Breaking Waves

Established in fluid dynamics terminology, "breaking wave" means the change of wave profile (shape) at the moment when the crest of the wave

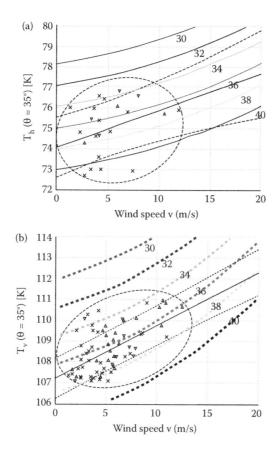

FIGURE 3.14
Radiation-wind dependencies and data comparisons at L band (1.4 GHz). (a) Horizontal polarization (solid); (b) vertical polarization (dash). Original experimental data—symbols, are from the WISE 2000. Model calculations: incidence 35°. $t = 15°C$; salinity is varied: $s = 30$–40 psu (marked). (Adapted from Camps A. et al. 2002. *IEEE Transactions on Geoscience and Remote Sensing.* 40(10):2117–2130.)

actually overturns. In view of electrodynamics and microwave remote sensing, this definition is not fully valid.

To compute emissivity, we have to characterize the wave breaking process by two different factors: geometrical and volume (Section 3.1.2). The first describes deformations of the surface geometry and the second describes the air–water mixing process, that is, phase transformation of the air–sea interface, under the influence of collapsing surface waves.

An option is to apply a combined microwave emission model considering the impact from both geometrical and volume factors statistically. Analogical task has been formulated and realized numerically for the analysis of high-resolution radar observations of breaking waves (Raizer 2013). The same approach can be used for passive microwave radiometry as well.

The averaged brightness temperature is written as a sum of independent contributions

$$T_B(\lambda, p) = \left[q_{sur} \kappa_{sur}(\lambda, p) + q_{vol} \kappa_{vol}(\lambda, p) \right] T_0, \quad q_{sur} + q_{vol} = 1, \quad (3.32)$$

where κ_{sur} and κ_{vol} are emissivities related to the surface and volume nonuniformities; q_{sur} and q_{vol} are the corresponding weight coefficients (area fractions); $p = h$, v (horizontal and vertical polarization, respectively); and T_0 is the thermodynamic temperature.

Spectral and polarization dependencies of emissivities $\kappa_{sur}(\lambda, p)$ and $\kappa_{vol}(\lambda, p)$ are calculated using available electromagnetic models (which are discussed in this section). Weight coefficients q_{sur} and q_{vol} are functions of the sea surface state parameters (wind speed, foam/whitecap coverage, boundary-layer characteristics). Indeed, it is an apparently simple microwave model because $T_B(\lambda, p)$ is a complicated nonlinear function of many geophysical variables.

3.5 Contributions from Foam, Whitecap, Bubbles, and Spray

Foam, whitecap, spray, and bubbles are the main factors contributing to ocean microwave emission at high wind speeds and strong gales. Microwave properties of these fascinating natural objects have been studied over the past 30 years by many researchers and scientists (including the author). Here, we discuss the most important data and results in an approachable manner with the goal to give the readers the best knowledge and understanding of the problem.

3.5.1 Microwave Properties of Foam and Whitecap

The first attempts to explain microwave emission effects induced by sea foam were undertaken in the 1970s. The study was initiated with pioneering remote sensing experiments (Nordberg et al. 1969; Webster et al. 1976). During the time, several microwave emission models of sea foam have been suggested. They are the following: two-phase air–water mixture (Droppleman 1970, Matveev 1971), multilayer structure of water and air films (Rozenkranz and Staelin 1972), and transitional dielectric layer and their combinations (Bordonskiy et al. 1978; Wilheit 1979; Raizer and Sharkov 1981). A number of numerical approximations of foam microwave emissivity were suggested as well (Stogryn 1972; Pandey and Kakar 1982). Although these approximations have certain limitations by the frequency range, they are still in use today (due to simplicity), for example, in global algorithms of satellite data assimilation (Kazumori et al. 2008).

In the late 1970s and early 1980s, the investigations of the microwave characteristics of foam and whitecap-like disperse structures were conducted in great detail; the results were published in a book by Cherny and Raizer (1998). These data have shown that microwave emission is defined not only by the void fraction of the foam medium (as many think) but also by the diffraction properties of individual bubbles and/or their aggregates. It was shown that a single bubble at millimeter and centimeter wavelengths represent *backbody-like* diffraction object. This effect is most pronounced at d ~ λ, where d is the size of the bubble.

Indeed, strong variations of the brightness temperature are observed in a wide range of λ = 0.3–18 cm depending on bubble parameters, stability, geometry, and concentration of foam. Earlier laboratory and nature microwave studies (Williams 1971; Vorsin et al. 1984; Smith 1988) have confirmed this statement as well. Below, we discuss selected fundamental experimental and model data.

3.5.1.1 Earlier Experimental Data

A set of one-channel radiometers operated at wavelengths λ = 0.26, 0.86, 2.08, 8, and 18 cm with a fluctuation sensitivity of 0.1–0.2 K was used in test measurements (Bordonskiy et al. 1978; Militskii et al. 1978). The goal of these experiments was to investigate the brightness temperature variations induced by the structural transformation of a foam layer. For this, a thick (~1–2 cm thickness) layer of chemical foam with polyhedral cells was created on the smooth water surface. The thick layer distracted gradually and after some time it was transformed into a stable emulsion monolayer of bubbles. The dynamics of a foam layer were registered by radiometers continuously.

During these test experiments, the following important results were obtained:

1. Multifrequency spectral dependencies of the microwave emission are defined by the thickness and disperse microstructure of a foam layer significantly.

2. At wavelengths λ = 0.26–8 cm, emissivity dominates due to a thin (~0.1 cm thickness) monolayer of bubbles located on the air–water interface.

3. At wavelengths λ = 0.26–2 cm, the emissivity of a foam layer of ~1–2 cm thickness is about 1, that is, a thick layer of foam represents an absolute black body.

4. At wavelengths λ = 0.26 and 0.8 cm, the emissivity of a foam layer is independent of polarization.

The microwave scattering characteristics of foam were also investigated in the laboratory using a bistatic reflection method (Militskii et al. 1977). The measurements of the scattering indicatrix were made at frequencies of 9.8, 36.2, and 69.9 GHz.

Active bistatic measurements have demonstrated that the electromagnetic properties of liquid foam are very close to that of a blackbody that is associated with a high absorption of microwaves by bubbles and polyhedral cells of foam. At the same time, it was found that a thin (~0.1 cm) emulsion monolayer of bubbles located on the water surface represents a two-dimensional diffraction grating with a spatial period $\Lambda \sim \lambda$. Selective reflection is explained by the resonance properties of effective bubble nodes that float on the water surface and form a monolayer. There is some analogy between selective reflection from bubbles and the diffraction of x-rays at crystal lattices or polyatomic molecules of liquids.

3.5.1.2 Model Data

The results from laboratory and field experiments clearly show that the theory of *heterogeneous mixtures* is unable to describe the microwave properties of foam and whitecap adequately in a wide range of microwave frequencies $\lambda = 0.3$–8 cm and two polarizations simultaneously. On the other hand, a common electromagnetic theory of wave propagation in dense media containing closely packed particles developed in Tsang et al. (1985, 2000a,b), Ishimaru (1991), and Apresyan and Kravtsov (1996) seems to be complicated for practical remote sensing applications on a regular basis. Moreover, it is not obvious that this theory can describe the emissivity of dynamic sea foam/whitecap with variable microstructure characteristics.

A practically usable variant of the model has been developed using the classical Lorentz–Lorenz equation (Raizer and Sharkov 1981; Dombrovsky and Rayzer 1992; Raizer 1992; Cherny and Raizer 1998) and/or analytic theory of thermal radiation in isotropic scattering media (Dombrovsky 1979; Dombrovsky and Baillis 2010).

The main parameter of this macroscopic model is an effective complex permittivity of polydisperse system of spherical bubbles. A single bubble at microwave frequencies is modeled by a hollow spherical particle with a thin water shell (Dombrovsky and Rayzer 1992). Scattering and absorption characteristics of a single bubble and the bubble system are computed using the Mie formulas modified for two-layer spherical particles.

The effective complex permittivity of the system is computed using the modified Lorentz–Lorenz formula

$$\varepsilon_{N\alpha} = \frac{1 + \frac{8}{3}\pi \overline{N\alpha}}{1 - \frac{4}{3}\pi \overline{N\alpha}}, \quad \overline{N\alpha} = \frac{k\int \alpha(a)f(a)da}{\frac{4}{3}\pi\int a^3 f(a)da}, \tag{3.33}$$

$$\alpha = a^3 \frac{(\varepsilon_0 - 1)(2\varepsilon_0 + 1)(1 - q^3)}{(\varepsilon_0 + 2)(2\varepsilon_0 + 1)(1 - q^3) + 9\varepsilon_0 q^3},$$

or using the Hulst (van de Hulst 1957) equation:

$$\varepsilon_{NS} = 1 + i4\pi \left(\frac{2\pi}{\lambda}\right)^{-3} \overline{NS_0}, \tag{3.34}$$

$$\overline{NS_0} = \frac{k \int S_0(a) f(a) da}{\frac{4}{3}\pi \int a^3 f(a) da}, \quad S_0 = \sum_{n=1}^{\infty} \frac{2n+1}{2}(A_n + B_n),$$

where $\varepsilon_{N\alpha}$ and ε_{NS} are complex effective permittivities of the polydisperse system of bubbles; $f(a)$ is the normalized size distribution function of bubbles; a is the external radius of a single bubble; δ is the thickness of the shell; N is the volume concentration of bubbles; k is the packing coefficient of bubbles; ε_0 is the complex permittivity of the shell medium (usually it is salt water); α is the complex polarizability of a single bubble; S_0 is the complex amplitude of the scattering "forward" by a single bubble; $q = 1 - \delta/a$ is the bubble's "filling" factor; and A_n and B_n are the complex Mie coefficients for a hollow spherical particle.

The first formula (3.33) takes into account dipole–dipole interaction of bubbles in a closely packed disperse system. The second formula (3.34) describes the contribution of the multipole moment (forward scattering) of noninteracting bubbles to the effective permittivity of the system. Both models operate with diffraction characteristics of bubbles.

A special numerical analysis by the Mie theory (Dombrovsky 1981) has shown that at wavelengths $\lambda = 0.2$–0.8 cm, large bubbles with a diameter $d = 0.1$–0.2 cm represent resonant objects with strong absorption and scattering. But at $\lambda = 2$–8 cm, spherical bubbles with diameter $d < 0.2$ cm are the Rayleigh particles. For such bubbles, the absorption cross sections essentially exceed the scattering cross section (Figure 3.15).

The resonance properties of bubbles cause an increase of both the real and imaginary parts of the complex effective permittivity $\varepsilon_{N\alpha}$ and ε_{NS} resulting in a change of full electromagnetic losses in disperse media. This broadband dielectric increment due to resonance effects yields more realistic spectral dependencies of the emissivity at a wide range of wavelengths $\lambda = 0.2$–8 cm.

The quasi-static macroscopic model (3.33) and (3.34) is based on a fundamental physical law modified for polydisperse media of spherical particles. Unlike mixing dielectric models, this model provides computing spectral dependencies of foam/whitecap emissivity adequately in a wide range of wavelengths $\lambda = 1.5$–21 cm. The critical parameters of the model are size distribution function $f(a)$ and the thickness of water shell δ of a bubble; these parameters should be specified properly.

More perspective models (Raizer 2006, 2007) involve vertical profiles of effective complex permittivity $\varepsilon_{N\alpha}(z)$ or $\varepsilon_{NS}(z)$, where z is the depth of a foam layer (Figure 3.16). These profiles depend on the vertical stratification of the

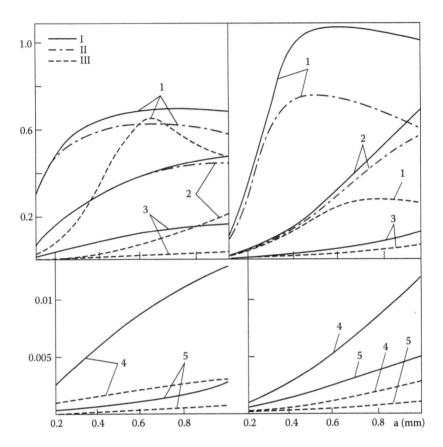

FIGURE 3.15

Effective factors of (I) extinction, (II) absorption, and (III) asymmetry of scattering versus outer radius of the bubble. Calculations using the Mie formulas for a hollow spherical particle. Thickness of water shell: (left panel) $\delta = 0.001$ cm; (right panel) $\delta = 0.005$ cm. Electromagnetic wavelength: (1) $\lambda = 0.26$ cm; (2) $\lambda = 0.86$ cm; (3) $\lambda = 2.08$ cm; (4) $\lambda = 8$ cm; (5) $\lambda = 18$ cm. (Adapted from Dombrovsky L. A. 1979. *Izvestiya, Atmospheric and Oceanic Physics.* 15(3):193–198 (translated from Russian); Dombrovsky L. A. 1981. *Izvestiya, Atmospheric and Oceanic Physics.* 17(3):324–329 (translated from Russian).)

phase components (water and air) and/or vertical distribution of the bubble's size in disperse medium. In the simplest case of a flat surface, spectral emissivity can be estimated using the standard formula $\kappa_f(\lambda) = 1 - |r_f(\lambda)|^2$, where $r_f(\lambda)$ is the spectral reflection coefficient of nonuniform medium with profiles $\varepsilon_{N\alpha}(z)$ or $\varepsilon_{NS}(z)$. Calculations of emissivity are made using a multilayer recursion algorithm operated with the complex Fresnel reflection coefficient (Raizer et al. 1986; Cherny and Raizer 1998).

Actually, such a stratified electromagnetic model describes a structure hierarchy and microwave properties of nonuniform sea foam/whitecap more realistically; it also provides a good agreement with accurate test radiometric measurements (Padmanabhan et al. 2007).

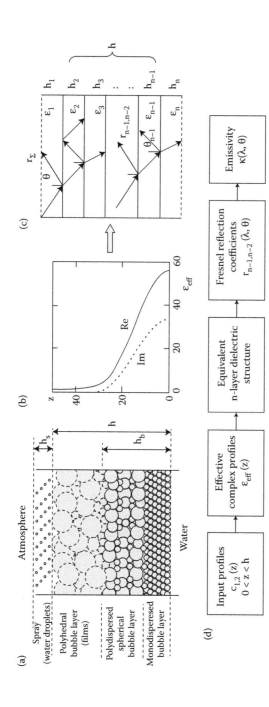

FIGURE 3.16

Structural microwave model of sea foam and multilayer simulation technique. (a) Hierarchical representation. (b) Continuous effective complex dielectric profile. (c) Multilayer dielectric model. (d) Flow chart for calculations of emissivity. (Adapted from Raizer V. 2006. *Proceedings of International Geoscience and Remote Sensing Symposium*, pp. 3672–3675. Doi: 10.1109/IGARSS.2006.941; Raizer V. 2007. *IEEE Transactions on Geoscience and Remote Sensing*. 45(10):3138–3144. Doi: 10.1109/TGRS.2007.895981.)

3.5.1.3 Recent Studies

Follow-up works concerning the contribution of foam/whitecap to the sea surface microwave emission were motivated mostly by upcoming space-based observations and hurricane forecasting programs. In this connection, the following studies have been performed over the last decade:

1. More detailed theoretical analysis and validation of mixing dielectric models for microwave radiometry (Anguelova 2008; Anguelova et al. 2009; Anguelova and Gaiser 2011, 2012, 2013; Hwang 2012; Wei 2013)

2. Creation of dense media wave propagation models based on numerical solution of Maxwell's equations (the so-called quasi-crystalline approximation), application of radiative transfer equation, and Monte Carlo simulations (Tsang et al. 2000a,b; Guo et al. 2001; Chen et al. 2003; Wei 2011)

3. Development of combined multidisperse macroscopic model for high-resolution multiband microwave radiometry and imagery (Raizer 2005, 2006, 2007, 2008)

4. More accurate measurements of foam/whitecap emissivity at different conditions (Rose et al. 2002; Padmanabhan and Reising 2003; Aziz et al. 2005; Salisbury et al. 2014; Wei et al. 2014a,b; Potter et al. 2015)

As a whole, the listed theoretical and experimental studies have led to progress in ocean microwave radiometry and data analyses. In particular, the emissivity of foam/whitecap at the selected microwave frequencies and view angles has been refined. It was also found experimentally that angular variations of the brightness temperature depend on foam structural characteristics.

Popular and relatively simple microwave models of foam/whitecap are based on the theory of heterogeneous dielectric mixtures. The macroscopic theory includes ~10+ different mixing formulas (Tinga et al. 1973; De Loor 1983; Sihvola 1999; Kärkkäinen et al. 2000). All these models operate with volume concentrations of phase components but, eventually, they do not describe the diffraction properties of internal microstructure elements. The mixing dielectric models may yield correct spectral values of foam emissivity at selected microwave frequencies under certain observation conditions. Specifically, the modified Lorentz–Lorenz formula (3.33) at $\delta = a$, $q = 0$ (i.e., when spherical bubble shell forms into water droplet) reduces to some dielectric mixing formulas.

Scattering and adsorption of microwaves lead to additional electromagnetic losses that affect the spectral and polarization characteristics of foam emissivity. An environmental example is a highly dynamic *whitecap bubble plume* (using Monahan's terminology), which produces contrast radio-brightness signatures even at S and L bands. The plume is two-phase turbulent flow of different particles of variable geometry and size. In this case, strong variations in emissivity are defined by two factors: void fraction and diffraction losses due to scattering by microstructure of internal particles (bubbles, droplets, and their aggregates).

The physical sense of all suggested foam/whitecap microwave models is the same: it is considered a two-phase disperse medium containing a small amount of water and a large amount of air. Such a composition always yields a high level of emissivity approximately in the range ~0.8–1.0 (depending on microwave frequency and view angle) due to a huge difference in the dielectric constant of water and air. The problem is in correct physical parameterization and experimental verification of the chosen models. The most important *criteria of adequateness* of the model is the ability to describe simultaneously emissivity in a wide spectral range of wavelengths from 0.3 to 30 cm and two polarizations, but not only at specified frequency bands.

Spectral dependencies of foam emissivity are summarized in Figure 3.17. The microwave model with effective dielectric profiles (Figure 3.16) yields a good agreement with the existing experimental data for the thickness of a foam layer h = 0.2–1.0 cm. Note that many experimental data have been collected earlier in the laboratory and from shipborne gyro-stabilizing platforms with improved spatial resolutions. In these cases, the best match

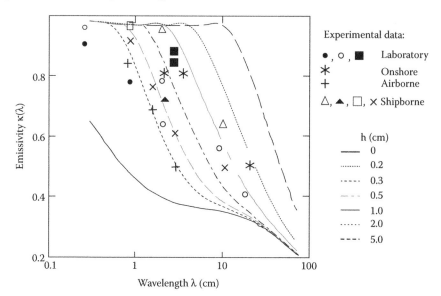

FIGURE 3.17
Spectral dependencies of foam emissivity (calculated at nadir). Model calculations (as in Figure 3.16) are shown for different thicknesses of foam layer h (marked). Experimental data are shown at different environmental conditions (unspecified). (Cherny I. V. and Raizer V. Yu. *Passive Microwave Remote Sensing of Oceans.* 195 p. 1998. Copyright Wiley-VCH Verlag GmbH & Co. KGaA; Raizer V. 2007. *IEEE Transactions on Geoscience and Remote Sensing.* 45(10):3138–3144. Doi: 10.1109/TGRS.2007.895981.)

can be realized through the correct parameterization of the model. Some deviations between model and experimental data will always occur due to foam microstructure variety and dynamical errors.

From Figure 3.17 it follows that the spectral dependencies of emissivity are divided well enough by the thickness of nonuniform foam layers. This may offer practical possibilities for passive microwave diagnostics foam/white-cap coverage on the ocean surface.

Most adequate microwave models of foam/whitecap for ocean remote sensing can be formulated and developed on the basis of the vector wave equation

$$\Delta \vec{E} + k^2 \vec{E} + \mathrm{grad}\left(\vec{E}\frac{1}{\varepsilon}\mathrm{grad}\varepsilon\right) = 0, \quad k(\vec{r}) = \frac{\omega^2}{c^2}\varepsilon(\vec{r}), \qquad (3.35)$$

where $\varepsilon(\vec{r}) = <\varepsilon> + \Delta\varepsilon(\vec{r})$ is the random field of complex dielectric permittivity, $<\varepsilon>$ is the mean of the permittivity, and $\Delta\varepsilon(\vec{r})$ is the fluctuation parts, dependent on the spatial coordinates $\vec{r} = \{x, y, z\}$; $k(\vec{r})$ is the propagation constant; and ω is the frequency of the electromagnetic wave.

The task can be divided into two parts. In the case of a vertically stratified medium, a basic solution of Equation 3.35 gives us the scattering and emission coefficients dependent on the profile $\varepsilon(z) = <\varepsilon> + \Delta\varepsilon(z)$. In the case of a horizontal distribution $\varepsilon(x, y) = <\varepsilon> + \Delta\varepsilon(x, y)$, the solution of Equation 3.35 can describe the spatial variations of the brightness temperature $\Delta T_B(x, y)$ associated with the horizontal nonuniformity of the foam/whitecap coverage (and microstructure).

Finally, there is a possibility to apply fractal-based formalism for describing the propagation (scattering and absorption) of microwaves in stochastic cluster disperse systems (for example, Babenko et al. 2003). The fractal dimension is a statistical parameter that provides a connection between electromagnetic and microstructure properties of disperse system. Fractal-based models are more flexible and compact, mathematically; they may describe adequately microwave characteristics of powerful two-phase disperse flows covering large spaces of the ocean at very high winds and hurricanes.

3.5.2 Emissivity of Spray

The first oceanographic considerations (Tang 1974; Barber and Wu 1997) have shown that "spray"—a mixture of water and air—yields an additional contribution to the sea surface brightness temperature. The estimates were made using the Fresnel reflection coefficients and empirical parameterization of the brightness temperature dependent on wind speed. Although the choice of such a picture of sea spray is a circuitous idea, the result was positive that has initiated further more detailed experimental studies. For example, recent data obtained from the Floating Instrument Platform (Savelyev et al. 2014) demonstrate a possibility to predict spray aerosol fluxes by measuring the sea surface brightness temperature at high winds. Satellite data also show that sea spray aerosol contributes to air–sea exchange (Anguelova and Webster 2006; Monique et al. 2010).

The important and, perhaps, the first detailed experimental studies of the droplet's flow and natural oceanic spray induced by breaking waves, were conducted in the mid-1980s using one-channel radiometer/scatterometer operated at the wavelength $\lambda = 0.8$ cm. These data have been published and discussed earlier in two books (Cherny and Raizer 1998; Sharkov 2007). Here, we briefly outline some important results that have a fundamental meaning for ocean remote sensing.

In laboratory experiments, it was found that variations of backscattering, extinction, and brightness temperature are defined by structural parameters and volume concentration of the droplet's flow. In the case of a low concentration ($c \leq 0.1\%$), the flow behaves like a discrete scattering and absorption medium containing small water droplets. In the case of a high concentration ($c \geq 4.5\%$), the flow behaves like a continuous turbulent media of closely spaced interacting droplets. These test experiments have provided a physical basis for further electromagnetic modeling of spray and its influence on ocean microwave emission.

Microwave observations from the ship have confirmed the statement that natural sea spray and foam/whitecap are absolutely different media in electromagnetic sense. In particular, time series measurements of wave breaking process conducted using radiometer-scatterometer at $\lambda = 0.8$ cm demonstrate the separate effects induced by foam and spray.

The results published in Cherny and Raizer (1998) show that backscattering cross section $\sigma_{bs}(t)$ and brightness temperature $T_B(t)$ vary with antiphase regime during the time. The value σ_{bs} increases faster than the value T_B. It means that scatterometric signal is more sensitive to spray injection, and radiometric signal is more sensitive to foam/whitecap production. The observed effect of *desynchronization* of backscattering and emission is explained by dissimilarity in the microstructure, spatial and temporal properties of sea spray and foam.

The impact of the near-surface sea spray and dense aerosol on ocean emissivity is defined not only by spray/aerosol coverage statistics (which is, probably, unknown in actual percentage) but also by electromagnetic properties of spray as a system of water droplets and their aggregates. Diffraction characteristics of spherical water droplets are specified using the Rayleigh/Mie scattering theory. The corresponding formulas, approximations, and numerical data are reported in some books (van de Hulst 1957; Deirmendjian 1969).

In the case of low-concentrated droplet's flow or spray, the radiative transfer theory can be applied to compute the increment of emissivity induced by spray. In the case of high-concentrated spray media, macroscopic mixing models are available to provide simple estimates. For example, effective complex dielectric permittivity of fine-dispersed spray or dense aerosol can be defined using the Maxwell–Garnett effective medium approximation related to the configuration "water-in-air" but not "air-in-water."

As mentioned above, sea spray can be modeled more adequately by a discrete system of spherical water droplets of different size and concentration. Direct numerical calculations of radiation characteristics of spray media at microwave frequencies were made first using the Mie theory (Dombrovsky and Rayzer 1992; Cherny and Raizer 1998). Model descriptions and theoretical results are considered below.

For a polydisperse system of water droplets, the volume factors of absorption $\bar{Q}_a$, scattering $\bar{Q}_s$, and attenuation $\bar{Q}_t$ are introduced:

$$\{\bar{Q}_a, \bar{Q}_s, \bar{Q}_t\} = \frac{3}{4}\frac{\varpi}{\rho}\int\{Q_a, Q_s, Q_t\}r^2 p(r)dr \Big/ \int r^3 p(r)dr, \qquad (3.36)$$

where Q_a and Q_s are the effective factors of absorption and scattering; $Q_t = Q_a + Q_s$ is the effective factor of attenuation for the water droplet with the radius r; ϖ is the mass concentration of the water; and ρ is the density of water. In microwave applications, the following two-parameter gamma distribution for spray droplet's size is used:

$$p(r) = \frac{A^{B+1}}{\Gamma(B+1)}r^B \exp(-Ar), \qquad (3.37)$$

where A and B are parameters. The "tail" of the distribution is sensitive to changes of wind speed and spray-generated conditions (Chapter 2).

Dimensionless factors of $Q_a(x)$, $Q_s(x)$, and $Q_t(x)$ are calculated using the Mie theory for the spherical particle (water droplet), where $x = 2\pi r/\lambda$ is the diffraction parameter. Note that all these factors are the function of the complex permittivity of the droplet's liquid (through the Mie complex coefficients), that is, they are dependent on electromagnetic wavelength, temperature, and salinity of seawater. There is a great sensitivity of spray scattering and absorption characteristics to parameters of spherical water droplets. Figure 3.18 demonstrates several numerical examples.

According to our calculations, the main electromagnetic properties of water droplets are as follows:

- In the microwave range $\lambda > 0.6$ cm, small-size droplets (with the radius $r < 0.05$ cm) are the particles with the Rayleigh law of scattering.
- Resonance region of scattering and absorption for large-size water droplets (with the radius $0.05 < r < 0.2$ cm) are manifested in the wavelengths range of $\lambda = 0.3$–5 cm.
- In the wavelengths range of $\lambda = 0.2$–8 cm, the radiation properties of the droplets depend on the water temperature.

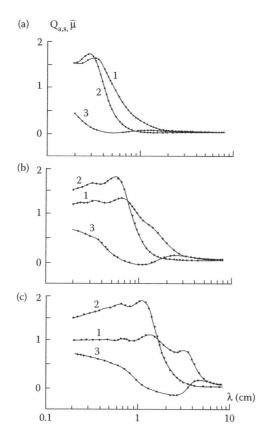

FIGURE 3.18

The effective factors of absorption (1), scattering (2), and asymmetry of scattering (3) calculated using the Mie formulas for spherical water droplets with different radius: (a) r = 0.05 cm; (b) r = 0.1 cm; (c) r = 0.2 cm. (Cherny I. V. and Raizer V. Yu. *Passive Microwave Remote Sensing of Oceans*. 195 p. 1998. Copyright Wiley-VCH Verlag GmbH & Co. KGaA.)

The emissivity induced by spray is determined not only by the diffraction properties of individual water droplets and their polydisperse system, but also by the *surface mass concentration* of water ωh in the spray layer of thickness h. Estimations were made using the analytic solution of the scalar radiative transfer equation (Cherny and Raizer 1998). The full theory of thermal radiation in disperse media is reported in a book by Dombrovsky and Baillis (2010).

Figure 3.19a shows the spectral dependencies of hemispherical emissivity $\kappa(\lambda)$ of the spray-water system for the case of a monodisperse layer of droplets with a different surface mass concentration $\omega h = 0.1$ and 0.01 g/cm^2. For large-size droplets (radius of the droplets is r = 0.2 cm) the curves $\kappa(\lambda)$ are reminiscent of the spectral dependencies $Q_a(\lambda)$. The scattering in the Rayleigh's region is weak compared with the absorption.

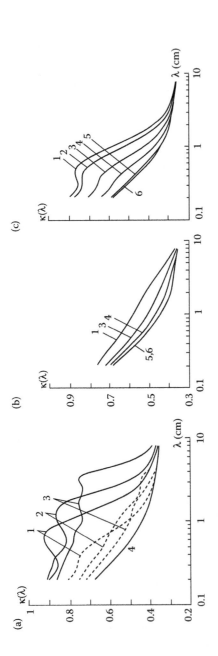

FIGURE 3.19

Emissivity spectrum of spray. Calculations: (a) monodispersed layer with variable droplet's radius: (1) $r = 0.05$ cm; (2) $r = 0.1$ cm; (3) $r = 0.2$ cm; (4) smooth water surface. Solid line: $\varpi h = 0.1$ g/cm^2. Dashed line: $\varpi h = 0.01$ g/cm^2. (b) and (c) Polydispersed layer with variable surface mass concentration: (1) $\varpi h = 0.1$ g/cm^2; (2) $\varpi h = 0.08$ g/cm^2; (3) $\varpi h = 0.05$ g/cm^2; (4) $\varpi h = 0.03$ g/cm^2; (5) $\varpi h = 0.01$ g/cm^2; (6) smooth water surface. (b) Small-size spray ($r_{max} \approx 0.01$ cm) and (c) large-size spray ($r_{max} \approx 0.1$ cm). (Cherny I. V. and Raizer V. Yu. *Passive Microwave Remote Sensing of Oceans.* 195 p. 1998. Copyright Wiley-VCH Verlag GmbH & Co. KGaA)

Figure 3.19b and c shows analogous dependencies for the case of a polydisperse system of droplets with the different parameters of size distributions (3.37). Here, the spectra of emissivity $\kappa(\lambda)$ depend considerably on the size distribution of spray droplets.

Theoretical analysis has revealed major microwave effects induced by the spray layer located over a smooth water surface. As a whole, spray causes the increase of emissivity at wavelengths $\lambda = 0.2$–5 cm. Variations of emissivity are determined mostly by resonance properties of large-size droplets, that is, by the "tails" of the size distribution (3.37). In the case of monodisperse spray, resonance effects are pronounced, while for polydisperse spray, they are smoothed out. The higher the mass concentration of the water, the higher the emissivity of the spray-water system. As a whole, we can argue that dense spray aerosol is an important contributor to the sea surface emissivity.

3.5.3 Contribution from Bubbles

The bubble populations are considered here as an intervening stage between a thin stable foam layer and underwater bubble medium. Bubble populations are located directly on the air–water surface producing single bubble clusters or their groups (Chapter 2). Sometimes, the interference picture from the films of bubbles is observed. Mathematically, the surface configuration can be represented by an ensemble of small- or large-size hemispherical shells floating on the water surface.

At high-resolution microwave observations, both volume and surface scattering effects from bubble populations should be considered simultaneously. Direct electromagnetic solutions related to such a complex geometry (like floating bubbles on the surface) is complicated enough in order to perform numerical analysis. However, it is possible to simplify the task considering, for example, the statistical ensemble of randomly oriented curved thin water films of bubbles covering the surface. Microwave characteristics are defined using physical optics approximation. A model of a "bubble dipole" (by analogy with acoustics) located over the dielectric (water) surface can be considered as well. As a whole, we assume that microwave properties of bubble populations may be more pronounced than a foam monolayer resonant character due to the involvement of both dielectric and diffraction effects.

Another important type of two-phase media is clouds of underwater gaseous microbubbles. The mechanisms of aeration and bubble generation in the ocean are considered in Chapter 2. Possible microwave effects can be estimated easily using well-known mixing dielectric formulas. This is exactly the case when macroscopic models can be applied and tested.

Effective complex permittivity of air–water mixture containing a large number of small bubbles can be defined by the following formula (De Loor 1983):

$$\varepsilon_m = \varepsilon_w + c(\varepsilon_i - \varepsilon_w) \sum_{j=1}^{3} \frac{1}{1 + \left(\dfrac{\varepsilon_i}{\varepsilon^*} - 1 \right) A_j}, \tag{3.38}$$

where $\varepsilon_w(\lambda)$ is the complex permittivity of the water; $\varepsilon_i = 1$ is the permittivity of air (or any gas); and c is the volume concentration of bubbles in the water. Variable geometry of bubbles is described over the form factor A_j. For spheres, it is $\{A_1 = A_2 = A_3 = 1/3\}$; for needles, it is $\{A_1 = A_2 = 1/2, A_3 = 0\}$, and for disks, it is $\{A_1 = A_2 = 0, A_3 = 1\}$. Formula (3.38) contains an unknown parameter ε^*, so-called effective internal dielectric constant, describing electrostatic contributions from other inclusions (in our case, bubbles) into effective constant of the mixture ε_m. It is assumed that the value of ε^* lies between two constant: $\varepsilon^* = \varepsilon_w$ and $\varepsilon^* = \varepsilon_m$. Substituting these constant into Equation 3.38, we obtain two boundaries for function $\varepsilon_m(c)$.

Formula 3.38 can be rearranged to give the real and imaginary parts of the complex permittivity $\varepsilon_m = \varepsilon'_m + i\varepsilon''_m$:

$$\varepsilon'_m = \varepsilon_{m\infty} + \frac{\varepsilon_{m0} - \varepsilon_{m\infty}}{1 + \left(\dfrac{\lambda_{ms}}{\lambda} \right)^2}, \quad \varepsilon''_m = \frac{\varepsilon_{m0} - \varepsilon_{m\infty}}{1 + \left(\dfrac{\lambda_{ms}}{\lambda} \right)^2} \cdot \frac{\lambda_{ms}}{\lambda}. \tag{3.39}$$

These relations describe the effective permittivity of a two-phase bubble medium as a liquid dielectric with parameters of relaxation: ε_{m0}, $\varepsilon_{m\infty}$, λ_{ms} analogously to Debye's equations for dielectric constants of water.

Figure 3.20 shows the Cole–Cole diagrams $\varepsilon''_m(\varepsilon'_m)$ calculated for the air bubbles-in-water medium. The values of air volume concentration (void fraction) $c = 0.05$–0.10 are close to the environmental range. All diagrams are designed using formula (3.39) when the electromagnetic wavelength λ is changed quietly from 0.1 cm to 30 cm. The value $c = 0$ corresponds to the case of the air-free water. The following features can be identified from these diagrams:

- The dielectric properties of the bubble medium are changed with the increase in the concentration of bubbles in the mixture. Cole–Cole diagrams are shifted to a lower value of the effective permittivity.
- In the case of freshwater, the Cole–Cole diagram holds its shape, but in the case of salt water, the shape is disrupted. In the wavelength range of $\lambda = 10$–30 cm, the right part of the diagram is streamed up (effect of salinity).

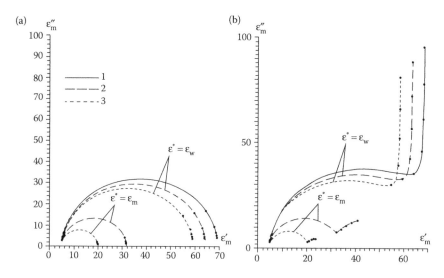

FIGURE 3.20
Complex effective permittivity of the bubbles-in-water emulsion. The air concentration (void fraction) is varied: (1) c = 0; (2) c = 0.05; (3) c = 0.1. (a) Fresh water; (b) salt water. (Cherny I. V. and Raizer V. Yu. *Passive Microwave Remote Sensing of Oceans*. 195 p. 1998. Copyright Wiley-VCH Verlag GmbH & Co. KGaA.)

- In the wavelength range of $\lambda = 0.1$–1 cm, the influence of the concentration of bubbles, temperature, and salinity of the water on the effective permittivity is weak.

The change of the dielectric properties of the two-phase skin layer produces considerable variations of microwave emission. In the case of a flat water surface, the brightness temperature of two-phase medium with the complex permittivity ε_m is equal to $T_B = (1 - |R(\varepsilon_m)|^2)T_0$, where R is the complex Fresnel reflection coefficient and T_0 is the thermodynamic temperature.

Spectral dependencies of the brightness temperature $T_B(\lambda)$ are shown in Figure 3.21. Calculations are made for both limiting cases: $\varepsilon^* = \varepsilon_0$; $\{A_1 = A_2 = A_3 = 1/3\}$ and $\varepsilon^* = \varepsilon_m$; $\{A_1 = A_2 = 0, A_3 = 1\}$. In the first case (bubbles are spheres), the dependence of the brightness temperature on the air concentration is low, but in the second case (bubbles are disks), the dependence is strong enough. It is important that microwave effects from the bubble medium appear in a wide range of wavelengths $\lambda = 0.1$–10 cm. The greater the wavelength of emission, the greater the value of the brightness temperature contrast. In fact, spectral dependencies $T_B(\lambda)$ reflect primarily a variance of the concentration of bubbles in the subsurface ocean layer. In the wavelength range $\lambda = 8$–21 cm, variations of the brightness temperature due to bubble aeration of the water reach about $\Delta T_B = 10$–15 K.

More detailed theoretical analysis was done using a two-phase bubble-layer model (Raizer 2004). These calculations based on statistical mixing

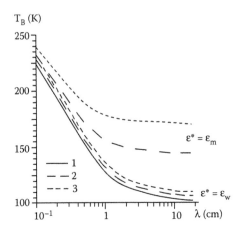

FIGURE 3.21
Brightness temperature of the bubble-in-water emulsion (at nadir). The air concentration (void fraction) is varied: (1) $c = 0$; (2) $c = 0.05$; (3) $c = 0.1$. (Cherny I. V. and Raizer V. Yu. *Passive Microwave Remote Sensing of Oceans.* 195 p. 1998. Copyright Wiley-VCH Verlag GmbH & Co. KGaA. Reproduced with permission.)

dielectric formula (Odelevskiy 1951) have demonstrated a great possibility for nonacoustic detection of gaseous bubbles at the subsurface layer of the ocean. In particular, at C, S, and L bands, there is the extreme sensitivity of microwave radiances to parameters of two-phase aeration flows such as bubbly jets and/or bubble wakes.

Bubbles near a sea surface can form different geometrical patterns: vortexes, stripes, spots, films; they change very rapidly; therefore, it is necessary to provide dynamic observations. For example, active–passive remote sensing experiments (Bulatov et al. 2003) show that "bubble signatures" can be detected and recognized very well by joint variations of microwave emission and backscatter registered simultaneously by radiometer and scatterometer. The Doppler spectrum variance can also be an indicator of bubble productivity at the near surface layer of the ocean.

3.5.4 Combined Foam–Spray–Bubbles Models

Sometimes, radiometer/scatterometer observations of a stormy ocean surface yield both short-term spatial and temporal variations of microwave emission/backscatter signals associated with wave breaking processes (fields). Microwave fluctuations occur not only under the influence of complex and variable geometry of the wave breaking crests but also due to joint impacts from subsurface bubbles, foam/whitecap, and spray as well. Evaluation of these particular effects can be done using combined electromagnetic disperse models, which have been developed and applied for analyses radiometer (Cherny and Raizer 1998; Raizer 1992, 2005, 2006, 2007) and radar (Raizer 2012, 2013) ocean remotely sensed data.

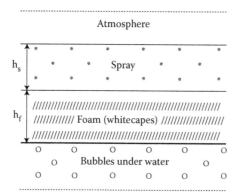

FIGURE 3.22
A combined microwave model of two-phase disperse medium at the ocean–atmosphere interface. (Cherny I. V. and Raizer V. Yu. *Passive Microwave Remote Sensing of Oceans.* 195 p. 1998. Copyright Wiley-VCH Verlag GmbH & Co. KGaA. Reproduced with permission.)

A combined electromagnetic model comprising different types of disperse media is shown schematically in Figure 3.22. In the common case, a three-layer system is considered: the upper first layer, bordering with the atmosphere, is the spray layer; the second layer is a foam (or whitecap); the third layer is a population of subsurface bubbles; below the uniform water medium is located. The system is characterized by a set of the following parameters:

- Temperature and salinity of the water
- Size distribution of spray droplets
- Water content of the spray or volume concentration of the water
- Height of the spray layer (h_s)
- Size distribution of foam/whitecap bubbles
- Average thickness of the bubble water shell
- Bulk bubble concentration in the foam/whitecap layer
- Thickness of the foam/whitecap layer (h_f)
- Void fraction (concentration) of gaseous bubbles-in-water at the upper ocean

The numerical algorithm is based on a combination of the scalar radiative transfer theory applied for discrete scattering system of water droplets (spray), the macroscopic theory of closely packed bubbles (foam/whitecap), and the models of dielectric mixtures (bubble-in-water populations). The composition (Figure 3.22) can yield unusual or even unpredictable variations of microwave emission depending on the set of chosen parameters. Moreover, it is clear that multiband measurements only will enable to distinguish effects induced by different disperse media by spectral and polarization characteristics of microwave emission.

Emissivity of the system can be calculated using the following radiative transfer equation:

$$\kappa(\lambda,\theta) \approx [1-r(\lambda,\theta)]\cdot\exp(-\tau_s/\cos\theta)+(1-\varpi)\cdot[1-\exp(-\tau_s/\cos\theta)]+$$
$$+ r(\lambda,\theta)\cdot(1-\varpi)\cdot[1-\exp(-\tau_s/\cos\theta)]\cdot\exp(-\tau_s/\cos\theta), \tag{3.40}$$

$$\tau_s = h_s\rho_s \int \pi a^2 Q_e(a)p(a)da, \tag{3.41}$$

$$Q_e(a) = \frac{2}{x^2}\sum_{n=1}^{\infty}(2n+1)\,\mathrm{Re}(a_n+b_n),$$

where $r(\lambda,\theta)$ is the power reflection coefficient of the water surface as a function of wavelength λ and incident angle θ; τ_s is the integral optical thickness of a spray layer; ϖ is the spectral albedo of spray droplets; ρ_s is the number of droplets in cubic centimeter (cm^{-3}); $h_s(cm)$ is the thickness of a spray layer; and $p(a)(cm^{-1})$ is the size distribution of spray droplets (usually it is the gamma distribution).

Scattering and absorption effects are incorporated into Equations 3.40 and 3.41 through the dimensionless extinction factor for a single water droplet $Q_e(a)$ of radius a, which is defined by the Mie complex coefficients a_n, b_n or through the Rayleigh approximation (at diffraction parameter $x \ll 1$, $x = 2\pi a/\lambda$). It is important to specify correctly the power Fresnel reflection coefficient $r(\lambda,\theta)$ in Equation 3.40. For example, for stratified macroscopic model of foam/whitecap, the reflection coefficient can be computed using the layer recursion technique (Section 3.3.5); for a smooth water surface, reflection and emission coefficients are defined using the Fresnel formulas (3.3) and (3.4).

Formula (3.40) can be simplified significantly if one considers a discrete medium of spherical particles (droplets or bubbles) and neglects the scattering term taking into account the absorption term only. In this case, spectral albedo $\varpi = 0$ and we obtain

$$\kappa(\lambda,\theta) = 1 - r(\lambda,\theta)\exp(-2\tau_0/\cos\theta), \tag{3.42}$$

where τ_0 is the integral optical thickness of disperse layer (foam, spray, or both of them). Formula (3.42) is valid at low τ_0 that is satisfied at wavelengths $\lambda > 3–5$ cm. It is a convenient formula for calculations of spectral dependencies of emissivity $\kappa(\lambda)$, especially at S and L bands. However, it does not describe polarization characteristics at $\lambda < 3$ cm.

Let us consider the following typical situations in more detail.

3.5.4.1 Spray over the Smooth Water Surface (Spray + Water)

The spray located above a smooth water surface always yields positive brightness temperature contrast. The effects depend not only on the diffraction properties of droplets but also on the mass concentration of the spray. The emissivity of the spray becomes higher with increasing mass concentration. The spectrum of microwave emission is more sensitive to variations of size distribution and the concentration of spray. In the case of monodisperse spray (the size of droplets is the same), resonance effects are pronounced, while for polydisperse spray (the size of droplets is different), the resonance effects are smoothed out.

Spray over a foam-free sea surface always yields positive brightness temperature contrast in a wide range of microwave wavelengths.

3.5.4.2 Spray over the Foam Coverage (Spray + Foam + Water)

If the foam/whitecap layer is incorporated into the combined model, spectral and polarization characteristics of microwave emission are changed considerably. The foam bubbles cause a strong absorption of microwave radiance, providing effects such as a "black body," and the spray causes both scattering and absorption effects. This complicated interplay yields nonmonotonic spectral dependencies of microwave emission with sigh-variable brightness temperature contrast. Moreover, the foam layer provides a substantial reduction of polarization differences.

Calculations of microwave emission show great sensitivity of microwave emission to parameters of spray (Cherny and Raizer 1998; Raizer 2007). Figure 3.23 shows numerical examples. Changes in the droplet size distribution affect the absolute value of brightness temperature contrast $\Delta T_B(\lambda)$ mostly at wavelengths of $\lambda = 0.3$–0.8 cm. Contrast $\Delta T_B(\lambda)$ is defined by the thickness of the intermediate layer of foam h. Positive contrasts $\Delta T_B(\lambda) > 0$ occur when spray is located over the water surface. If the spray is located over any foam/whitecap surface, negative contrasts $\Delta T_B(\lambda) < 0$ can appear within the range of $\lambda = 0.3$–8 cm, depending on the incidence angle and polarization. In this case, we observe the so-called cooling effect induced by the spray itself. The positive contrast is a result of the absorption, and the negative contrast is a result of the scattering of microwave radiance on water droplets.

The spray over a foam layer on the sea surface yields as positive as negative brightness temperature contrast depending on the thickness and properties of the foam layer.

3.5.4.3 Influence of Bubbles Populations (Spray + Foam + Subsurface Bubbles + Water)

A layer of gaseous bubbles located below foam (or foam + spray) layers can also give some minor changes in the spectral dependencies of microwave emission. Some "calming" effect in microwave emission occurs at a wide

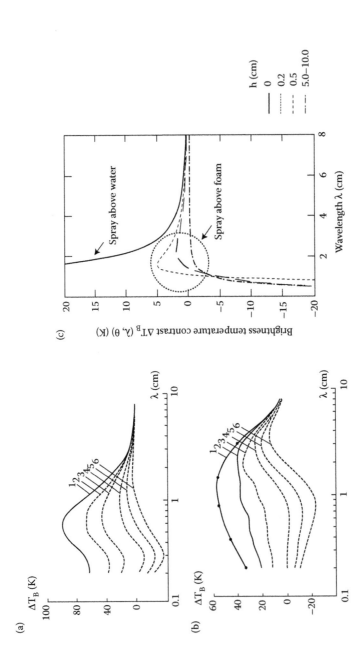

FIGURE 3.23
Brightness temperature contrast of the water–foam–spray system due to spray impact *only* (at nadir). Thickness $h = h_f$ of the foam layer is varied: (1) spray over free-foam water surface ($h = 0$). Spray over foam surface: (2) $h = 0.5$ cm; (3) $h = 0.1$ cm; (4) $h = 0.2$ cm; (5) $h = 0.3$ cm; (6) $h = 0.5$ cm. (a) Small-size ($r_{max} \approx 0.01$ cm) spray; (b) large-size ($r_{max} \approx 0.1$ cm) spray; (c) "cooling" and "warming" microwave effects induced by spray. Thickness of foam layer h is varied (marked). Thickness of spray layer is constant $h_s = 10$ cm. (Cherny I. V. and Raizer V. Yu. *Passive Microwave Remote Sensing of Oceans*. 195 p. 1998. Copyright Wiley-VCH Verlag GmbH & Co. KGaA; Adapted from Raizer V. 2007. *IEEE Transactions on Geoscience and Remote Sensing*. 45(10):3138–3144. Doi: 10.1109/TGRS.2007.895981.)

range of microwave frequencies. As a whole, increases of the bubble concentration in the water decreases variations of microwave emission that is assassinated with large penetration depth of microwave inside the two-phase aeration medium.

3.6 Measuring the (Sub)Surface Turbulence

Some literature data have shown that ocean surface turbulence can be observed using accessible remote sensing methods: radar (SAR) (George and Tatnall 2012), infrared (IR) instrumentation (Veron et al. 2009), and high-resolution optical imagery (Keeler et al. 2005; Gibson et al. 2008). But there is no experimental evidence that a passive microwave radiometer is capable of observing directly (sub)surface turbulence or turbulence intermittency.

Meanwhile, strong wave–turbulence interactions and coupling effects can change the parameters of the air–sea interface, which, in principle, gives us a good chance for passive microwave diagnostics of small-scale (~1–100 m) and/or even fine-scale (~0.1–1 m) ocean turbulence. Although this is a challenging task, some theoretical aspects of the problem could be outlined.

Supposedly, variations of the brightness temperature induced by small-scale (sub)surface turbulence are defined by two factors: (1) short-term fluctuations of surface roughness and (2) changes of the electromagnetic skin layer properties due to mixing. In both situations, the emissivity should be a function of dynamic characteristics of the air–sea interface (including surface roughness) associated with the turbulence regime.

3.6.1 Surface Turbulence

Surface turbulence represents stochastic pulsations of a surface fluid producing roughness instabilities, patches, vortexes, and/or specific patterns. To describe a spatial field of radio-brightness fluctuations, we introduce by analogy with turbulence in the atmosphere (Tatarskii 1961; Kutuza 2003) the second-order structure function of the brightness temperature

$$D_{T_B}(\vec{r}) = <[\tilde{T}_B(\vec{r}_1 + \vec{r}) - \tilde{T}_B(\vec{r}_1)]^2>, \tag{3.43}$$

where $\tilde{T}_B(\vec{r}_1) = T_B(\vec{r}_1) - \overline{T}_B(\vec{r}_1)$ is a fluctuation part of the radio-brightness field $T_B(\vec{r})$ at the point $\vec{r}_1$ and $\overline{T}_B(\vec{r})$ is its mean part at the point $\vec{r}_1$. Supposedly, the square root $\Delta T_B = \sqrt{D_{T_B}}$ is a measure of the intensity of brightness temperature fluctuations.

On the other hand, we can use the resonance model (3.19) in order to define a fluctuation part of the brightness temperature (relative to unperturbed turbulence-free surface)

$$\tilde{T}_B(\vec{r}) \approx 2T_0 k_0^2 \iint G(K, k_0; \varphi)\tilde{F}(K, \varphi; \vec{r})KdKd\varphi \approx T_0 k_0^2 B(k_0)\tilde{A}(\vec{r}), \quad (3.44)$$

where $B(k_0)$ is a coefficient obtained from the integration of the resonance function $G(K, k_0)$ at $K = k_0$; $\tilde{A}(\vec{r}) = A(\vec{r}) - A_0$. In the convolution (3.44), a fluctuation part of the surface wave number spectrum is $\tilde{F}(K, \vec{r}) = F(K, \vec{r}) - F_0(K)$, where $F(K, \vec{r})$ and $F_0(K)$ are the perturbed (due to turbulence) and unperturbed (turbulence-free) wave number spectrum, respectively. The spectrum is written in general form as $F(K, \vec{r}) = A(\vec{r})K^{-n}$ and $F_0(K) = A_0K^{-n}$ (A_0 and n = const).

The combination of Equations 3.43 and 3.44 gives an analytical structure function

$$D'_{T_B} = <[\tilde{T}_B(\vec{r}_1 + r) - \tilde{T}_B(\vec{r})]^2> \approx [2T_0 k_0^2 B(k_0)]^2 \cdot <[\tilde{A}(\vec{r}_1 + \vec{r}) - \tilde{A}(\vec{r})]^2>, \quad (3.45)$$

and from Equation 3.45, we obtain

$$D'_{T_B}(\vec{r}) \approx \alpha_0 \cdot [2T_0 k_0^2 B_0(k_0)]^2 \cdot D_\xi(\vec{r}), \quad (3.46)$$

where

$$D_\xi(\vec{r}) = <[\tilde{\xi}(\vec{r}_1 + \vec{r}) - \tilde{\xi}(\vec{r}_1)]^2> = 2\int(1 - \cos(-K \cdot \vec{r}))F(K, \vec{r})dK$$

is the structure function of roughness elevations; $\tilde{\xi}(\vec{r}) = \xi(\vec{r}) - \xi_0(\vec{r})$ is the surface elevation increment; $\tilde{\xi} = \alpha_0 \tilde{A}$, and α_0 is some coefficient.

We may believe now that small-scale surface turbulence can be observed by a passive microwave radiometer through the measurement of structure function of radio-brightness $D_{T_B}(\vec{r})$. An important suggestion here is that both structure functions of velocity fluctuations $D_u(\vec{r}) = <[\tilde{u}(\vec{r}_1 + \vec{r}) - \tilde{u}(\vec{r}_1)]^2>$ and surface elevations $D_\xi(\vec{r}) = <[\tilde{\xi}(\vec{r}_1 + \vec{r}) - \tilde{\xi}(\vec{r}_1)]^2>$ in the presence of turbulence are close to each other. For example, $D_\xi(\vec{r}) \sim D_u(\vec{r}) \approx (\varepsilon_r r)^{\xi_n}$, $r = |\vec{r}|$. If this is true, then the value of scaling exponent ξ_n can be defined from the experiment. Note that the radio-brightness structure function $D'_{T_B}(\vec{r})$ is a function of microwave frequency.

3.6.2 Subsurface Turbulence

Here, we consider the possibilities for microwave diagnostics of subsurface fine-scale turbulence. This type of turbulence can exist in the form of turbulent spots, thermohaline intrusions, aerating jets (bubbly flows), or other mixed substance. Turbulent features can occur under the influence of a number of factors: mixing, thermohaline convection, breaking of internal waves, collapsing turbulent wakes, cavitation, bubble activity, and other causes (Chapter 2).

In the *electrodynamic sense*, the ocean mixing environment can be represented by a composite medium with changing dielectric and physical parameters. In this case, microwave radiometry is able to provide diagnostics of turbulent features located at a thin (<1 m) subsurface layer. The technique is based on the method of microwave impedance spectroscopy. Impedance spectroscopy (also called dielectric spectroscopy) is widely used in electrical engineering and antenna technology for measuring parameters of composition dielectric materials as a function of electromagnetic frequency (Kremer and Schönhals 2003; Barsoukov and Macdonald 2005).

Observations of subsurface turbulence can be conducted using a multifrequency set of passive microwave radiometers operating at C, S, and L bands ($\lambda = 4–30$ cm). At these microwave bands, there is a good sensitivity of radiometric signals to critical ocean parameters and small-scale surface nonuniformities due to the increased penetration depth $\ell \sim \lambda/(2\pi\sqrt{|\hat{\varepsilon}|})$, where $\hat{\varepsilon}$ is the complex permittivity. The value ℓ varies depending on the electromagnetic wavelength and structural and physical properties of a mixed medium.

The technique is based on the evaluation of effective complex impedance $\hat{Z}_{\text{eff}}(\lambda)$ and/or effective complex dielectric permittivity $\hat{\varepsilon}_{\text{eff}}(\lambda)$ of a thin upper ocean layer through multifrequency polarimetric measurements of brightness temperature $T_B(\lambda,\theta;p)$ at selected electromagnetic wavelengths (for example, $\lambda = 4, 6, 8, 10, 18, 21, 30$ cm), specified view angles θ, and polarizations ($p = h, v$).

In the case of nadir observations ($\theta = 0$), complex impedance $\hat{Z}_{\text{eff}}(\lambda)$ and permittivity $\hat{\varepsilon}_{\text{eff}}(\lambda)$ can be retrieved from the measured brightness temperature $T_B(\lambda)$ using the following well-known relationship:

$$\frac{T_B(\lambda)}{T_0} = \kappa(\lambda) = \left[1 - \left|\hat{R}_{\text{eff}}(\lambda)\right|^2\right] = \left[1 - \left|\frac{\hat{Z}_{\text{eff}}(\lambda) - Z_0}{\hat{Z}_{\text{eff}}(\lambda) + Z_0}\right|^2\right], \quad (3.47)$$

where $\hat{R}_{\text{eff}}(\lambda)$ is the effective complex reflection coefficient, $\kappa(\lambda)$ is the emission coefficient, $\hat{Z}_{\text{eff}}(\lambda) = \sqrt{\mu_{\text{eff}}/\hat{\varepsilon}_{\text{eff}}(\lambda)}$ (for nonmagnetic medium $\mu_{\text{eff}} = 1$), $Z_0 = \sqrt{\mu_0/\varepsilon_0}$ is the wave impedance of free space, and T_0 is the thermodynamic temperature.

For normalized complex impedance

$$\frac{\hat{Z}_{\text{eff}}}{Z_0} = r + jx = \frac{1 + \hat{R}_{\text{eff}}}{1 - \hat{R}_{\text{eff}}}, \quad \hat{R}_{\text{eff}} = u + jv, \quad |\hat{R}_{\text{eff}}|^2 = u^2 + v^2, \quad (3.48)$$

we obtain real and imaginary parts of normalized impedance related to the Smith chart of transmission line

$$r = \frac{1 - u^2 - v^2}{(1-u)^2 + v^2} \quad \text{and} \quad x = \frac{2v}{(1-u)^2 + v^2}. \tag{3.49}$$

Effective complex permittivity is defined as

$$\hat{\varepsilon}_{eff} = \varepsilon_r + j\varepsilon_i = \left[\frac{1 + u + jv}{1 - u - jv}\right]^2. \tag{3.50}$$

Impedance-based approach allowing the use of the Fresnel formulas for effective reflection coefficient $|\hat{R}_{eff}|^2 = u^2 + v^2$ that at nadir viewing geometry yields (3.50). Moreover, at wavelengths $\lambda > 4$–6 cm for uniform seawater environment $u \gg v$; therefore, Equations 3.49 and 3.50 can be reduced to

$$r \approx \frac{1 - u^2}{(1-u)^2} = \frac{1+u}{1-u} \quad \text{and} \quad \hat{\varepsilon}_{eff} \approx \varepsilon_r = \left[\frac{1+u}{1-u}\right]^2, \tag{3.51}$$

where $u = \sqrt{|\hat{R}_{eff}|^2} = \sqrt{1 - \kappa}$.

In common case, when both real and imaginary parts of complex reflection coefficient $\hat{R}_{eff} = u + jv$ should be defined from the brightness temperature, the retrieval procedure is more complicated than technique (3.49) through (3.51). It is required to use two-position off-nadir multifrequency polarimetric measurements.

In Figure 3.24, we present the suggested methodology of dielectric spectroscopy and a flowchart for the retrieval of $\hat{\varepsilon}_{eff}(\lambda)$ from microwave radiometric measurements. The algorithm operates with the relationships between the complex Fresnel coefficients at two polarizations (Azzam 1979, 1986; Shestopaloff 2011).

On the other hand, the changes in the complex permittivity of a mixed medium can be computed, for example, using the Havriliak–Negami Equation 3.14 and available mixing dielectric formulas (Tinga et al. 1973; De Loor 1983; Sihvola 1999; Kärkkäinen et al. 2000). Some results are demonstrated below.

Figure 3.25 shows several examples of diagrams $\text{Im}\{\hat{\varepsilon}_{eff}(\lambda)\}$ versus $\text{Re}\{\hat{\varepsilon}_{eff}(\lambda)\}$ computed for different situations called "gas-in-liquid" intrusions and "liquid-in-liquid" intrusions. To provide their numerical detailization and evaluation the best, the electromagnetic wavelength is varied in a wide range $\lambda = 0.3$–30 cm with a very small discrete interval $\Delta\lambda = 0.1$ cm.

Calculations of the complex effective permittivity $\hat{\varepsilon}_{eff}(\lambda)$ are made using the *Wiener* matrix formula in the case of "liquid-in-liquid" intrusions and the (Odelevskiy 1951) statistical mixing formula in case of "gas-in-liquid"

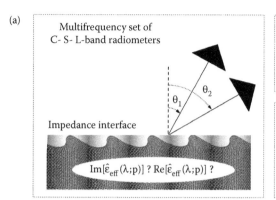

(a)

(b)

FIGURE 3.24
Dielectric spectroscopy of a mixed ocean medium. (a) Experimental methodology. (b) Flow chart and basic formulas for the retrieval of the effective complex permittivity of subsurface layer from multiband radiometric measurements.

intrusions. These diagrams demonstrate differences between dielectric spectra obtained for different types of a mixed medium. *Distinctivity* of the diagrams is a graphically measurable property providing a physical basis for the implementation of impedance microwave spectroscopy.

In the case of "gas-in-liquid" intrusions, significant variations in the dielectric spectra are observed due to a large penetration depth of microwaves into a mixed aerated medium. The behavior of the diagrams considerably depends on the void fraction of gaseous intrusions.

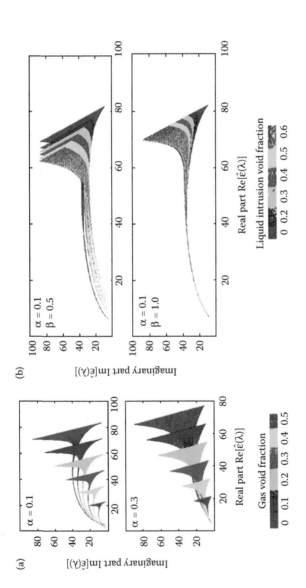

FIGURE 3.25

Effective dielectric spectra computed for two types of subsurface mixed turbulent medium. Parameters of the *Havriliak–Negami* equation (α, β) are marked. (a) Left panel: Gas-in-liquid intrusions (like "cavitation flow" and/or "bubbly jet"). This case is assigned to "detection performance level." (b) Right panel: Liquid-in-liquid intrusions (like "turbulent spots" and/or "salt fingers"). This case is assigned to "detection difficulty level." In both panels (a) and (b), color scale (is not a bar) shows the value of the void fraction for the corresponding type of intrusions.

In the case of "liquid-in-liquid" intrusions, more compact localization of dielectric spectra is observed. It is assumed that an ambient seawater and liquid intrusions have different physical properties. Important regulating parameters here are exponents α and β in the Havriliak–Negami Equation 3.14. The volume concentration of liquid intrusions is also a variable parameter. The microwave response is weak but, probably, measurable using sensitive low-frequency radiometers.

All dielectric spectra presented in Figure 3.25 are different. Frequency shifting and the value of the effective complex permittivity depend on a combination of input parameters in the dielectric mixing model. However, there is a certain tendency in these results. In the case of "gas-in-liquid" intrusions, dielectric spectra are distinguished very well at both high- and low-frequency microwave bands; but in the case of "liquid-in-liquid" intrusions, dielectric *distinctivity* may exist at low-frequency bands only. Therefore, we believe that relevant information can be obtained mostly at C, S, and L bands.

As a whole, an impedance-based concept developed here has potential capabilities for microwave diagnostics of weakly emergent natural phenomena, including subsurface turbulence and mixing processes.

3.7 Emissivity of Oil Spills and Pollutions

Remote sensing methods have already been progressively established for the environmental monitoring of oil spills in the ocean. Along with radar, infrared, lidar, and video technologies, passive microwave radiometry is an efficient tool for the aerial surveillance of oil pollutions. For example, microwave scanning radiometer can determine an oil spill layer thickness between 50 μm and 3 mm.

The capability of microwave radiometer is defined by the high sensitivity of spectral emissivity to surface oil films due to a large difference between the dielectric properties of oil products and seawater. Earlier studies (Hollinger and Mennella 1973; Hurford 1986; Skou 1986; Lodge 1989; Krotikov et al. 2002) demonstrate that the electrophysical properties of the sea surface are significantly modified by oil films. In this context, it is important to separate the following two mechanisms affecting the sea microwave emission in the presence of oil slicks: (1) the roughness change due to strong attenuation of the high-frequency components of the wave spectrum, and (2) the change of electromagnetic wave propagation due to the action of the oil slick as an electromagnetic matching layer between free space and seawater.

The first mechanism has been considered in Section 2.5.7. The damping effect yields low-contrast variations of the brightness temperature $\Delta T_B \approx 2$–3 K in the presence of monomolecular films or very thin oil slicks on the sea surface. It is important to evaluate the second mechanism—the impedance matching—which provides high-contrast variations of brightness temperature up to $\Delta T_B \approx 60$ K depending on surface conditions. This effect is directly associated with the film thickness, type, and dielectric properties of oil products (gasoline, benzene, and petroleum).

3.7.1 Dielectric Properties of Oil and Derivatives

Some experimental data on the dielectric constant of the oils and their derivatives are summarized in Table 3.1. The value of the real part of the complex dielectric constant is equal to $\varepsilon' = 1.8 - 3.0$. Dielectric losses of the oil products are small. The tangent of dielectric losses is order to $\tan(\varepsilon'/\varepsilon'') = 10^{-3}$. Laboratory measurements show that ε' linearly depends on the specific weight of a purified oil and the time of its stay in open air. With the growth of temperature, the value ε' insignificantly subsides.

Dielectric properties of water-in-oil emulsion differ essentially from the dielectric properties of the pure or raw oil. For the calculation of the complex permittivity of the water-in-oil emulsion, mixing dielectric formulas, for example, Equation 3.38 can be employed. In this case, it is needed to change the parameters: $\varepsilon_w \rightarrow \varepsilon_i$ and $\varepsilon_i \rightarrow \varepsilon_w$, where ε_i and ε_w are complex dielectric constants of oil and water; and A_j is the form-factor of water inclusions. Calculations show that typical values of the effective permittivity of emulsion are $\varepsilon'_m = 1.5 - 5.0$ and $\varepsilon''_m = 1.0 - 3.5$ in the wavelength range $\lambda = 0.2$–2.0 cm at the bulk concentration of the water in emulsion $c = 0$–0.2. At the large concentrations $c > 0.3$–0.4, the influence of form-factor A_j on the value of ε_m is essential.

3.7.2 Microwave Model and Effects

A simple microwave model of the surface covered by emulsion layer is two-layer dielectric model. The emissivity of a film–water system κ_m is calculated through the Fresnel reflection coefficient of flat two-layer dielectric medium (Landau and Lifshitz 1984):

TABLE 3.1

Dielectric Parameters of Oil Products

Oil Product	Dielectric Constant (Real Part), ε'	Dielectric Loss Tangent, $\tan(\varepsilon'/\varepsilon'')$	Temperature (°C)	Frequency
Benzene	2.25–2.27	–	–	0.1–1 GHz; 5–10 GHz
Industrial benzene	2.10	3×10^{-3}	–	35 GHz
Oil raw	2.12–2.25	10^{-3}	20–30	10 MHz; 3.9–10 GHz
Oil distiller	1.8–3.0	5×10^{-3}	23	37 GHz

$$\kappa_m = 1 - \left| \frac{r_{12}e^{-2i\Psi} + r_{23}}{r_{12}r_{23} + e^{-2i\Psi}} \right|^2, \quad \Psi = \frac{2\pi h}{\lambda}\sqrt{\varepsilon_m - \sin^2\theta}, \tag{3.52}$$

where r_{12}, r_{23} are the coefficients of the reflection from the corresponding film boundaries; h is the thickness of a film; and θ is the angle of view. Effective complex permittivity $\varepsilon_m(\lambda;t,s)$ as a function of (λ) electromagnetic wavelength, (t) temperature, and (s) salinity of water is calculated by formula (3.38).

The brightness temperature contrast of a film-water system is $\Delta T_B = (\kappa_m - \kappa_0) T_0$, where κ_0 is the emissivity of the oil-free flat water surface. In the wavelength range of $\lambda = 0.2$–0.8 cm, contrast ΔT_B increases with the growth of the bulk concentration of water in emulsion, it can reach the value $\Delta T_B = 60$–80 K. At wavelengths $\lambda > 2$ cm, the contrast is $\Delta T_B < 2$ K.

Figure 3.26 illustrates typical interference dependencies of the brightness temperature of a two-layer dielectric system. The period of oscillations is estimated by

$$H = \frac{\lambda}{2\sqrt{\varepsilon'_m - \sin^2\theta}}. \tag{3.53}$$

In the case of a raw oil (without dielectric losses), the amplitude of oscillations is constant and is independent of the thickness of a film. But in the

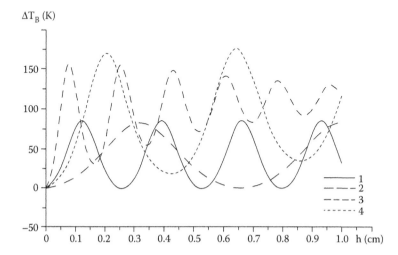

FIGURE 3.26
Dependence of the brightness contrast versus thickness of oil film. Modeling at the wavelengths $\lambda = 0.8$ and 2 cm (at nadir). Bulk concentration of the water in emulsion is varied: (1) c = 0 and $\lambda = 0.8$ cm; (2) c = 0 and $\lambda = 2$ cm; (3) c = 0.5 and $\lambda = 0.8$ cm; (4) c = 0.5 and $\lambda = 2$ cm. (Cherny I. V. and Raizer V. Yu. *Passive Microwave Remote Sensing of Oceans.* 195 p. 1998. Copyright Wiley-VCH Verlag GmbH & Co. KGaA.)

case of water-in-oil emulsion (the losses are introduced), the oscillations will attenuate, and the asymptotic level of the brightness temperature is determined by the complex dielectric constant of the film only.

Polarization dependencies reveal an interesting effect when microwave emission does not depend on the thickness of the dielectric layer. At the Bruster angle of view about $\theta = 65-68°$ and vertical polarization, the brightness temperature is defined by the dielectric constant of the layer only (Figure 3.27). It means that under these angles of view, it is possible to estimate the value of the dielectric constant of the oil product using polarization microwave radiometric measurements. At the grazing angles of view $\theta = 70-80°$ at the vertical polarization, the increase of bulk concentration of water in emulsion causes a decrease of the brightness temperature contrast.

Two-channel regression of the brightness contrasts due to the influence of the oil film, calculated for a pair of wavelengths $\lambda = 0.8$ and 2 cm, is shown in Figure 3.28. The curves have a form of loops and represent interference features of the reflection and emission from a two-layer dielectric medium. The reading of the oil thickness with the discrete of 50 μm is marked by the dots.

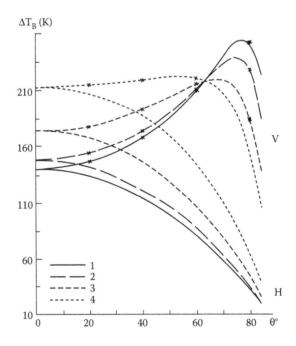

FIGURE 3.27
Brightness temperature of oil films versus view angle. Modeling at the wavelength $\lambda = 0.8$ cm. Polarizations: vertical (V) and horizontal (H). The bulk concentration of the water in emulsion is varied: $1 - c = 0$; $2 - c = 0.2$; $3 - c = 0.4$; $4 - c = 0.5$. (Cherny I. V. and Raizer V. Yu. *Passive Microwave Remote Sensing of Oceans.* 195 p. 1998. Copyright Wiley-VCH Verlag GmbH & Co. KGaA. Reproduced with permission.)

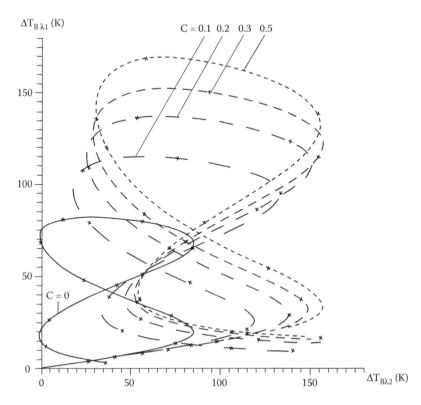

FIGURE 3.28

Two-channel regression of the brightness temperature contrasts due to oil film at the wavelengths $\lambda_1 = 2$ cm and $\lambda_2 = 0.8$ cm (at nadir). The bulk concentration of the water in emulsion is varied in the interval $c = 0...0.5$. (Cherny I. V. and Raizer V. Yu. *Passive Microwave Remote Sensing of Oceans*. 195 p. 1998. Copyright Wiley-VCH Verlag GmbH & Co. KGaA.)

The comparison of the theoretical and experimental data is shown on Figure 3.29. We use the results from airborne and laboratory microwave measurements of natural oil slicks. Variation of parameters of the model allows one to obtain the optimal agreement between theory and experiments. In the case of airborne measurements at the wavelengths $\lambda = 0.8$ and 2 cm, the best agreement occurs for the water-in-oil emulsion; in the case of laboratory measurements at the wavelength $\lambda = 2$ cm, it will be for the raw oil, spreading on the smooth water surface.

Using multifrequency microwave radiometry, it is possible to measure the thickness of oil slick and concentration of emulsions. The combination of radar and radiometer microwave data provides more detailed information concerning the behavior of spreading oil and its characteristics. Our and other studies show that the wavelength range of $\lambda = 0.3–2.0$ cm and view angles $\theta = 0–30°$ are the optimal observation parameters needed for microwave diagnostics of oil pollutions in the ocean.

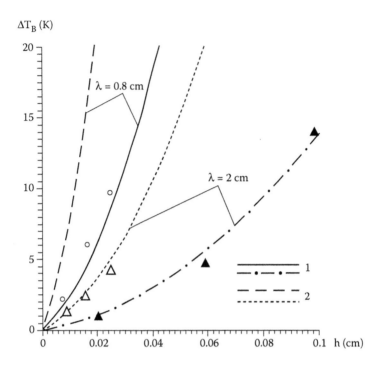

FIGURE 3.29
Dependence of the brightness temperature contrast on thickness of oil films at the wavelengths $\lambda = 0.8$ and 2 cm (at nadir). Calculations: (1) raw oil; (2) water-in-oil emulsion ($c = 0.5$). Experimental data: o, Δ—airborne; $\blacktriangle$—laboratory. (Cherny I. V. and Raizer V. Yu. *Passive Microwave Remote Sensing of Oceans*. 195 p. 1998. Copyright Wiley-VCH Verlag GmbH & Co. KGaA. Reproduced with permission.)

3.8 Influence of Atmosphere

In many applications, including radiometric observations from high-altitude platforms—aircraft or satellite—there is a need to estimate the influence of Earth's atmosphere on the ocean microwave emission. Under the assumption of horizontally uniform isotherm atmosphere, the total brightness temperature of the ocean–atmosphere system measured by radiometer can be expressed through a solution of the radiative transfer equation

$$T_B(\lambda, \theta; \vec{q}) = \kappa(\lambda, \theta; \vec{q}) \cdot T_0 \cdot \exp(-\tau_\lambda / \cos\theta)$$
$$+ T_{up}(\lambda, \theta) + r(\lambda, \theta; \vec{q}) \cdot [T_{down}(\lambda, \theta)$$
$$+ T_{cosm} \cdot \exp(-\tau_\lambda / \cos\theta)] \cdot \exp(-\tau_\lambda / \cos\theta), \qquad (3.54)$$

$$T_{up}(\lambda, \theta) = [1 - \exp(-\tau_\lambda / \cos\theta)] \cdot T_{a\uparrow}(\lambda), \qquad (3.55)$$

$$T_{down}(\lambda,\theta) = [1 - \exp(-\tau_\lambda / \cos\theta)] \cdot T_{a\downarrow}(\lambda). \tag{3.56}$$

In Equations 3.54 through 3.56, $\kappa(\lambda,\theta;\bar{q})$ and $r(\lambda,\theta;\bar{q})$ are coefficients of emission (emissivity) and reflection (power reflectivity) of the ocean surface; $T_{up}(\lambda,\theta)$ and $T_{down}(\lambda,\theta)$ are atmospheric upwelling and downwelling brightness temperatures; $T_{a\uparrow}(\lambda)$ and $T_{a\downarrow}(\lambda)$ are the corresponding atmospheric effective radiating temperatures; $T_{cosm} = 2.7\,K$ is space (galaxy) radiating temperature; $\tau_\lambda = \tau_{\lambda o} + \tau_{\lambda w} + \tau_{\lambda c}$ is the total atmospheric absorption related to oxygen ($\tau_{\lambda o}$), water vapor ($\tau_{\lambda w}$), and cloud liquid water ($\tau_{\lambda c}$); T_0 is the surface temperature. For a smooth water surface, $\kappa(\lambda,\theta;\bar{q}) = 1 - r(\lambda,\theta;\bar{q})$.

In Equation 3.54, the coefficients of emission $\kappa(\lambda,\theta;\bar{q})$ and reflection $r(\lambda,\theta;\bar{q})$ are functions of the wavelength λ, the incidence angle θ, polarization, and vector parameter $\bar{q}$, which is used for characterizations of ocean and atmosphere conditions. In particular, emissivity of the ocean surface $\kappa(\lambda,\theta;\bar{q})$ can be defined using a multifactor ocean surface model. On the other hand, radiative transfer Equation 3.54 through 3.56 is used for the retrieval of surface and atmospheric parameters from microwave radiometric data, for example (Guissard 1998; Mitnik and Mitnik 2003).

An accurate numerical modeling using Equation 3.54 requires *a priori* knowledge of atmospheric parameters. However, at high-resolution radiometric observations, atmospheric and surface microwave signatures can be distinguished by specific radio-brightness contrasts and/or their spectral and polarization characteristics. Therefore, the atmospheric terms in Equation 3.54 may not be so critical in some environmental situations (for example, at clear air and/or cloudless sky) although this question is important for data interpretation and deserves more attention.

3.9 Summary

In this chapter, the main mechanisms of the ocean microwave emission are considered. Microwave emission characteristics are defined by three factors: (1) dielectric properties of the seawater (its skin layer), (2) geometrical, and (3) volume nonuniformities of the air–sea interface.

Dielectric permittivity is a function of electromagnetic wavelength, temperature, and salinity of water. There is strong dispersion of the complex permittivity in millimeter and centimeter ranges of electromagnetic wavelengths. At centimeter wavelengths, the influence of temperature on the permittivity and emissivity is more pronounced; at decimeter wavelengths, the influence of salinity on emissivity is much stronger.

The contributions of geometrical and volume nonuniformities to ocean emissivity are quite different. Geometrical nonuniformities—surface waves, roughness, and turbulence—produce low-contrast brightness temperature signatures (usually up to 5 K at the near-nadir incident angles); volume nonuniformities—foam, whitecap, spray, bubbles, oil pollutions, and emulsions—produce high-contrast brightness temperature signatures (usually up to 20–30 K for real-world situations). In the general case, variations of ocean microwave emission measured by radiometer are defined by a joint impact from both geometrical and volume factors, which could be considered as statistical ensembles.

As a whole, microwave radiation of the ocean–atmosphere system is characterized by multiparameter spectral brightness temperature functional

$$T_{B\lambda}(\vec{q}) = F_\lambda(V, T, S, H; O_{2A}, W_A, Q_A), \qquad (3.57)$$

which involves wind speed (V), sea surface temperature (T), salinity (S), hydrodynamic response (H) associated with certain phenomenon or event; atmospheric oxygen (O_{2A}), water vapor (W_A), and cloud liquid water. An integral functional (3.57) provides a mathematical basis for the retrieval of a number of ocean and atmosphere parameters by multiband passive microwave radiometric measurements. In atmospheric remote sensing, this task can be considered using the mathematical method for solving incorrectly posed problems (Tikhonov and Arsenin 1977; Doicu et al. 2010).

References

Akhadov, Y. Y. 1980. *Dielectric Properties of Binary Solutions: A Data Handbook.* Pergamon, Oxford, England.

Akhmanov, S. A., D'yakov, Yu. E., and Chirkin, A. S. 1981. *Introduction to Statistical Radio Physics and Optics.* Nauka, Moscow (in Russian).

Anguelova, M. D. 2008. Complex dielectric constant of sea foam at microwave frequencies. *Journal of Geophyical Research.* 113(C08001). Doi: 10.1029/2007JC004212.

Anguelova, M. D. and Gaiser, P. W. 2011. Skin depth at microwave frequencies of sea foam layers with vertical profile of void fraction. *Journal of Geophysical Research.* 116(C11002). Doi: 10.1029/2011JC007372.

Anguelova, M. D. and Gaiser, P. W. 2012. Dielectric and radiative properties of sea foam at microwave frequencies: Conceptual understanding of foam emissivity. *Remote Sensing.* 4:1162–1189. Doi: 10.3390/rs4051162.

Anguelova, M. D. and Gaiser, P. W. 2013. Microwave emissivity of sea foam layers with vertically inhomogeneous dielectric properties. *Remote Sensing of Environment.* 139:81–96.

Anguelova, M. D., Gaiser, P. W., and Raizer, V. 2009. Foam emissivity models for microwave observations of oceans from space. In *Proceedings of International Geoscience and Remote Sensing Symposium,* 12–17 July 2009, Cape Town, South Africa, Vol. 2, pp. II-274–II-277. Doi: 10.1109/IGARSS.2009.5418061.

Anguelova, M. D. and Webster, F. 2006. Whitecap coverage from satellite measurements: A first step toward modeling the variability of oceanic whitecaps. *Journal of Geophysical Research.* 111(C03017):1–23.

Apresyan, L. A. and Kravtsov, Y. A. 1996. *Radiation Transfer: Statistical and Wave Aspects*. Gordon and Breach Publishers, Amsterdam.

Aziz, M. A., Reising, S. C., Asher, W. E., Rose, L. A., Gaiser, P. W., and Horgan, K. A. 2005. Effects of air–sea interaction parameters on ocean surface microwave emission at 10 and 37 GHz. *IEEE Transactions on Geoscience and Remote Sensing*. 43(8):1763–1774.

Azzam, R. M. A. 1979. Direct relation between Fresnel's interface reflection coefficients for the parallel and perpendicular polarizations. *Journal of the Optical Society of America*. 69(7):1007–1016.

Azzam, R. M. A. 1986. Relationship between the p and s Fresnel reflection coefficients of an interface independent of angle of incidence. *Journal of the Optical Society of America A*. 3(7):928–929.

Babenko, V. A., Astafyeva, L. G., and Kuzmin, V. N. 2003. *Electromagnetic Scattering in Disperse Media: Inhomogeneous and Anisotropic Particles*. Springer, Berlin.

Barber, Jr., R. B. and Wu, J. 1997. Sea brightness temperature and effects of spray and whitecaps. *Journal of Geophysical Research*. 102(C3):5823–5827.

Barsoukov, E. and Macdonald, J. R. 2005. *Impedance Spectroscopy: Theory, Experiment, and Applications*, 2nd edition. Wiley, Hoboken, NJ.

Basharinov, A. E., Gurvich, A. S., and Yegorov, S. T. 1974. *Radio Emission of the Earth as a Planet*. Moscow. Nauka (in Russian).

Bass, F. G. and Fuks, I. M. 1979. *Wave Scattering from Statistically Rough Surfaces*. Pergamon, Oxford, New York.

Belyaev, Y. K. and Nosko, V. P. 1969. Characteristics of excursions above a high level for a Gaussian process and its envelope. *Theory of Probability and Its Application*. 14(2):296–309.

Bettenhausen, M. H., Smith, C. K., Bevilacqua, R. M., Nai-Yu Wang, N.-Yu., Gaiser, P. W., and Cox, S. 2006. A nonlinear optimization algorithm for WindSat wind vector retrievals. *IEEE Transactions on Geoscience and Remote Sensing*. 44(3):597–610.

Bogorodskiy, V. V., Kozlov, A. I., and Tuchkov, L. T. 1977. *Radio Thermal Emission of the Earth's Covers*. Leningrad. Hydrometeoizdat (in Russian).

Bolotnikova, G. A., Irisov, V. G., Raizer, V. Yu., Smirnov, A. I., and Etkin, V. S. 1994. Variations of the natural emission of the ocean in the 8 and 18 cm bands. *Soviet Journal of Remote Sensing*. 11(3):393–404 (translated from Russian).

Bordonskiy, G. S., Vasil'kova, I. B., Veselov, V. M., Vorsin, N. N., Militskiy, Yu. A., Mirovskiy, V. G., Nikitin, V. V. et al. 1978. Spectral characteristics of the emissivity of foam formations. *Izvestiya, Atmospheric and Oceanic Physics*. 14(6):464–469 (translated from Russian).

Born, M. and Wolf, E. 1999. *Principles of Optics*, 7th edition. Cambridge University Press, Cambridge.

Brekhovskikh, L. 1980. *Waves in Layered Media*, 2nd edition. (Applied Mathematics and Mechanics, volume 16). Academic Press, New York.

Bulatov, M. G., Kravtsov, Yu. A., Pungin, V. G., Raev, M. D., and Skvortsov, E. I. 2003. Microwave radiation and backscatter of the sea surface perturbed by underwater gas bubble flow. In *Proceedings of International Geoscience and Remote Sensing Symposium*, 21–25 July 2003, Toulouse, France, Vol. 4, 2668–2670. Doi: 10.1109/IGARSS.2003.1294545.

Bunkin, F. V. and Gochelashvili, K. S. 1968. Bursts of a random scalar field. *Radiophysics and Quantum Electronics*. 11(12):1059–1063 (translated from Russian).

Camps, A., Corbella, I., Vall-Llossera, M., Duffo, N., Torres, F., Villarino, R., Enrique, L. et al. 2003. L-band sea surface emissivity: Preliminary results of the Wise-2000 campaign and its application to salinity retrieval in the SMOS mission. *Radio Science*. 38(4):MAR 36-1–MAR 36-8.

Camps, A., Font, J., Etcheto, J., Caselles, V., Weill, A., Corbella, I., Vall-Llosser, M. et al. 2002. Sea surface emissivity observations at L-band: First results of the wind and salinity experiment WISE-2000. *IEEE Transactions on Geoscience and Remote Sensing*. 40(10):2117–2130.

Camps, A., Font, J. Vall-llossera, M., Gabarro, C., Corbella, I., Duffo, N., Torres, F. et al. 2004. The WISE 2000 and 2001 field experiments in support of the SMOS mission: Sea surface L-band brightness temperature observations and their application to sea surface salinity retrieval. *IEEE Transactions on Geoscience and Remote Sensing*. 42(4):804–823.

Camps, A., Vall-llossera, M., Villarino, R., Reul, N., Chapron, B., Corbella, I., Duff, N. et al. 2005. The emissivity of foam covered water surface at L-band: Theoretical modeling and experimental results from the Frog 2003 field experiment. *IEEE Transactions on Geoscience and Remote Sensing*. 43(5):925–937.

Chen, D., Tsang, L., Zhou, L., Reising, S. C., Asher, W. E., Rose, L. A., and Ding, K. H. 2003. Microwave emission and scattering of foam based on Monte Carlo simulations of dense media. *IEEE Transactions on Geoscience and Remote Sensing*. 41(4):782–789.

Cherny, I. V. and Raizer, V. Yu. 1998. *Passive Microwave Remote Sensing of Oceans*. Wiley, Chichester, England.

Cole, K. S. and Cole, R. H. 1941. Dispersion and absorption in dielectrics I. Alternating current characteristics. *Journal of Chemical Physics*. 9:341–351. Doi: 10.1063/1.1750906.

Cole, K. S. and Cole, R. H. 1942. Dispersion and absorption in dielectrics II. Direct current characteristics. *Journal of Chemical Physics*. 10:98–105. Doi: 10.1063/1.1723677.

Cox, T. S. and Munk, W. 1954. Statistics of the sea surface derived from sun glitter. *Journal of Marine Research*. 13:198–227.

Davidson, D. W. and Cole, R. H. 1951. Dielectric relaxation in glycerol, propylene glycol, and n-propanol. *Journal of Chemical Physics*. 19(12):1484–1490.

Debye, P. 1929. *Polar Molecules*. Chemical Catalog, New York.

Deirmendjian, D. 1969. *Electromagnetic Scattering on Spherical Polydispersions*. American Elsevier Publishing Company, New York.

De Loor, G. P. 1983. The dielectric properties of wet materials. *IEEE Transactions on Geoscience and Remote Sensing*. 21(3):364–369.

Demir, M. A. and Johnson, J. T. 2007. Fourth-order small-slope theory of sea-surface brightness temperatures. *IEEE Transactions on Geoscience Electronics*. 45(1):175–186.

Doicu, A., Trautmann, T., and Schreier, F. 2010. *Numerical Regularization for Atmospheric Inverse Problems*. Springer Praxis, Berlin, Germany.

Dombrovsky, L. A. 1979. Calculation of thermal radio emission from foam on the sea surface. *Izvestiya, Atmospheric and Oceanic Physics*. 15(3):193–198 (translated from Russian).

Dombrovsky, L. A. 1981. Absorption and scattering of microwave radiation by spherical water shells. *Izvestiya, Atmospheric and Oceanic Physics*. 17(3):324–329 (translated from Russian).

Dombrovsky, L. A. and Baillis, D. 2010. *Thermal Radiation in Disperse Systems: An Engineering Approach.* Begell House Publishers Inc., Redding, CT.

Dombrovsky, L. A. and Rayzer, V. Yu. 1992. Microwave model of a two-phase medium at the ocean surface. *Izvestiya, Atmospheric and Oceanic Physics.* 28(8):650–656 (translated from Russian).

Droppleman, J. D. 1970. Apparent microwave emissivity of sea foam. *Journal of Geophysical Research.* 75(3):696–698.

Dzura, M. S., Etkin, V. S., Khrupin, A. S., Pospelov, M. N., and Raev, M. D. 1992. Radiometers-polarimeters: Principles of design and applications for sea surface microwave emission polarimetry. In *Proceedings of International Geoscience and Remote Sensing Symposium,* May 26–29, 1992, Houston, TX, Vol. 2, pp. 1432–1434. Doi: 10.1109/IGARSS.1992.578475.

Ellison, W. Balana, A., Delbos, G., Lamkaouchi, K., Eymard, L., Guillou, C., and Prigent, C. 1998. New permittivity measurements of seawater. *Radio Science.* 33(3):639–648.

Ellison, W. J., English, S. J., Lamkaouchi, K., Balana, A., Obligis, E., DeBlonde, G., Hewison, T. J., Bauer, P., Kelly, G., and Eymard, L. 2003. A comparison of ocean emissivity models using the Advanced Microwave Sounding Unit, the Special Sensor Microwave Imager, the TRMM Microwave Imager, and airborne radiometer observations. *Journal of Geophysical Research.* 108(D21):ACL 1.1–ACL 1.14. Doi: 10.1029/2002JD003213.

Etcheto, J., Dinnat, E. P., Boutin, J., Camps, A., Miller, J., Contardo, S., Wesson, J., Font, J., and Long, D. G. 2004. Wind speed effect on L-band brightness temperature inferred from EuroSTARRS and WISE 2001 field experiments. *IEEE Transactions on Geoscience and Remote Sensing.* 42(10):2206–2213.

Etkin, V. S., Raev, M. D., Bulatov, M. G., Militsky, Y. A., Smirnov, A. V., Raizer, V. Y., Trokhimovsky, Y. A. et al. 1991. *Radiohydrophysical AeroSpace Research of Ocean.* Academy of Science, Space Research Institute, Moscow, Russia, Technical Report. IIp-1749.

Etkin, V. S., Vorsin, N. N., Kravtsov, Y. A., Mirovskiy, V. G., Nikitin, V. V., Popov, A. E., and Troitskiy, I. A. 1978. Discovering critical phenomena under thermal microwave emission of the uneven water surface. *Radiophysics and Quantum Electronics.* 21(3):316–318 (translated from Russian).

Franceschetti, G., Imperatore, P., Iodice, A., Riccio, D., and Ruello, G. 2008. Scattering from layered structures with one rough interface: A unified formulation of perturbative solutions. *IEEE Transactions on Geoscience and Remote Sensing.* 46(6):1634–1643.

Fung, A. K. 1994. *Microwave Scattering and Emission Models and Their Applications.* Artech House, Norwood, MA.

Fung, A. K. and Chen, K.-S. 2010. *Microwave Scattering and Emission Models for Users.* Artech House, Norwood, MA.

Gadani, D. H., Rana, V. A., Bhatnagar, S. P., Prajapati, A. N., and Vyas, A. D. 2012. Effect of salinity on dielectric properties of water. *Indian Journal of Pure & Applied Physics.* 50:405–410.

George, S. G. and Tatnall, A. R. L. 2012. Measurement of turbulence in the oceanic mixed layer using Synthetic Aperture Radar (SAR). *Ocean Science Discussions.* 9(5):2851–2883. Doi: 10.5194/osd-9-2851-2012.

Gershenzon, V. E., Raizer, V. Yu., and Etkin, V. S. 1982. The transition layer method in the problem of thermal radiation from rough surface. *Radiophysics and Quantum Electronics.* 25(11):914–918 (translated from Russian).

Gibson, C. H., Bondur, V. G., Keeler, R. N., and Leung, P. T. 2008. Energetics of the beamed Zombie turbulence maser action mechanism for remote detection of submerged oceanic turbulence. *Journal of Applied Fluid Mechanics*. 1(1):11–42.

Goodberlet, M. A., Swift, C. T., and Wilkerson, J. C. 1989. Remote sensing of ocean surface wind with the Special Sensor Microwave/Imager. *Journal of Geophysical Research*. 94(C10):14547–14555.

Grankov, A. G. and Milshin, A. A. 2015. *Microwave Radiation of the Ocean-Atmosphere: Boundary Heat and Dynamic Interaction*, 2nd edition. Springer, Cham, Switzerland.

Guillou, C., Ellison, W., Eymard, L., Lamkaouchi, K., Prigent, C., Delbos, G., Balana, G. and Boukabara, S. A. 1998. Impact of new permittivity measurements on sea surface emissivity modeling in microwaves. *Radio Science*. 33(3):649–667. Doi: 10.1029/97RS02744.

Guissard, A. 1998. The retrieval of atmospheric water vapor and cloud liquid water over the oceans from a simple radiative transfer model: Application to SSM/I data. *IEEE Transactions on Geoscience and Remote Sensing*. 36(1):328–332.

Guo, J., Tsang, L., Asher, W., Ding, K.-H., and Chen, C.-T. 2001. Applications of dense media radiative transfer theory for passive microwave remote sensing of foam covered ocean. *IEEE Transactions on Geoscience and Remote Sensing*. 39(5):1019–1027.

Hasted, J. B. 1961. The dielectric properties of water. In *Progress in Dielectrics*, J. B. Birks and J. Hart (eds.), Vol. 3, pp. 103–149. Heywood, London.

Havriliak, S. and Negami, S. 1967. A complex plane representation of dielectric and mechanical relaxation processes in some polymers. *Polymer*. 8:161–210. Doi: 10.1016/0032-3861(67)90021-3.

Ho, W. and Hall, W. F. 1973. Measurements of the dielectric properties of sea water and NaCl solutions at 2.65 GHz. *Journal of Geophysical Research*. 78(27):6301–6315.

Hollinger, J. P. 1970. Passive microwave measurements of the sea surface. *Journal Geophysical Research*. 75(27):5209–5213. Doi: 10.1029/JC075i027p05209.

Hollinger, J. P. 1971. Passive microwave measurements of sea surface roughness. *IEEE Transactions on Geoscience Electronics*. 9(3):165–169.

Hollinger, J. P. and Mennella, R. A. 1973. Oil spills: Measurements of their distributions and volumes by multifrequency microwave radiometry. *Science*. 181:54–56.

Hollinger, J. P., Peirce, J. L., and Poe, G. A. 1990. SSM/I Instrument evaluation. *IEEE Transactions on Geoscience and Remote Sensing*. 28(5):781–790.

Hurford, N. 1986. Use of airborne microwave radiometry for the detection and investigation of oil slicks at sea. *Oil and Chemical Pollution*. 3(1):5–18. Doi: 10.1016/S0269-8579(86)80010-7.

Hwang, P. A. 2012. Foam and roughness effects on passive microwave remote sensing of the ocean. *IEEE Transactions on Geoscience and Remote Sensing*. 50(8):2978–2985.

Il'in, V. A., Kamenetskaya, M. S., Rayzer, V. Yu., Fatykhov, K. Z., and Filonovich, S. R. 1988. Radiophysical studies of nonlinear surface waves. *Izvestiya, Atmospheric and Oceanic Physics*. 24(6):467–471 (translated from Russian).

Il'in, V. A., Kasymov, S. S., Rayzer, V. Yu., Stepanishceva, M. N., and Fatykhov, K. Z. 1991. Laboratory studies of disturbances on a surface caused by falling rain. *Izvestiya, Atmospheric and Oceanic Physics*. 27(5):399–402 (translated from Russian).

Il'in, V. A., Naumov, A. A., Rayzer, V. Yu., Filonovich, S. R., and Etkin, V. S. 1985. Influence of short gravity waves on the thermal radiation from the surface of water. *Izvestiya, Atmospheric and Oceanic Physics*. 21(1):59–63 (translated from Russian).

Ilyin, V. A. and Raizer, V. Yu. 1992. Microwave observations of finite-amplitude water waves. *IEEE Transactions on Geoscience and Remote Sensing.* 30(1):189–192. Doi: 10.1109/36.124232.

Imperatore, P., Iodice, A., and Riccio, D. 2009. Electromagnetic wave scattering from layered structures with an arbitrary number of rough interfaces. *IEEE Transactions on Geoscience and Remote Sensing.* 47(4):1056–1072.

Irisov, V. G. 1987. PhD. Space Research Institute (IKI). Moscow. Russia (in Russian).

Irisov, V. 1994. Small-scale expansion for electromagnetic-wave diffraction on a rough surface. *Waves in Random Media.* 4(4):441–452.

Irisov, V. G. 1997. Small-slope expansion for thermal and reflected radiation from a rough surface. *Waves in Random Media.* 7(1):1–10.

Irisov, V. G. 2000. Azimuthal variations of the microwave radiation from a slightly non-Gaussian sea surface. *Radio Science.* 35(1):65–82.

Irisov, V. G., Trokhimovskii, Yu. G., and Etkin, V. S. 1987. Radiothermal spectroscopy of the ocean surface. *Soviet Physics, Doklady.* 32(11):914–915 (translated from Russian).

Ishimaru, A. 1991. *Electromagnetic Wave Propagation, Radiation, and Scattering.* Englewood Cliffs, Prentice Hall, NJ.

Ishimaru, V. G. 1991. Electromagnetic model for rough surface microwave emission and reconstruct ripple spectrum parameters. In *Proceedings of International Geoscience and Remote Sensing Symposium,* June 3–6, 1991, Helsinki, Finland, Vol. 3, pp. 1271–1273.

Jakeman, E. 1991. Non-Gaussian statistical models for scattering calculations. *Waves Random Media.* 3(1):S109–S119.

Janssen, M. A. 1993. *Atmospheric Remote Sensing by Microwave Radiometry.* John Wiley and Sons, New York, NY.

Johnson, J. T. 2005. A study of rough surface thermal emission and reflection using Voronovich's small slope approximation. *IEEE Transactions on Geoscience and Remote Sensing.* 43(2):306–314.

Johnson, J. T. 2006. An efficient two-scale model for the computation of thermal emission and atmospheric reflection from the sea surface. *IEEE Transactions on Geoscience and Remote Sensing.* 44(3):560–568.

Johnson, J. T., Kong, J. A., Shin, R. T., Staelin, D. H., O'Neill, K., and Lohanick, A. W. 1993. Third Stokes parameter emission from a periodic water surface. *IEEE Transactions on Geoscience and Remote Sensing.* 31(5):1066–1080.

Johnson, J. T., Kong, J. A., Shin, R. T., Yueh, S. H., Nghiem, S. V., and Kwok, R. 1994. Polarimetric thermal emission from rough ocean surfaces. *Journal of Electromagnetic Waves and Applications.* 8(1):43–59.

Johnson, J. T. and Zang, M. 1999. Theoretical study of the small slope approximation for ocean polarimetric thermal emission. *IEEE Transactions on Geoscience and Remote Sensing.* 37(5):2305–2316.

Joseph, G. 2005. *Fundamentals of Remote Sensing,* 2nd edition. University Press, (India) Private Limited, Hyderguda, Hyderabad, India.

Joshi, A. S. and Kurtadikar, M. L. 2013. Study of seawater permittivity models and laboratory validation at 5 GHz. *Journal of Geomatics. Indian Society of Geomatics.* 7(1):33–40.

Kärkkäinen, K. K., Sihvola, A. H., and Nikoskinen, K. I. 2000. Effective permittivity of mixtures: Numerical validation by the FDTD method. *IEEE Transactions on Geoscience and Remote Sensing.* 38(3):1303–1308.

Kazumori, M., Liu, Q., Treadon, R., and Derber, J. C. 2008. Impact study of AMSR-E radiances in the NCEP Global Data Assimilation System. *Monthly Weather Review.* 136(2):541–559. Doi: 10.1175/2007MWR2147.1.

Keeler, R. N., Bondur, V. G., and Gibson, C. H. 2005. Optical satellite imagery detection of internal wave effects from a submerged turbulent outfall in the stratified ocean. *Geophysical Research Letters*. 32(12):L12610. Doi: 10.1029/2005GL022390.

Kerr, Y. H., Waldteufel, P., Wigneron, J.-P., Martinuzzi, J.-M., Font, J., and Berger, M. 2001. Soil moisture retrieval from space: The Soil Moisture and Ocean Salinity (SMOS) mission. *IEEE Transactions on Geoscience and Remote Sensing*. 39(8):1729–1735.

Kirchhoff, G. 1860. Über das Verhaltnis zwischen dem Emissionsvermögen und dem Absorptionsvermögen. *der Körper für Wärme und Licht. Poggendorfs Annalen der Physik und Chemie*. 109:275–301. English translation by F. Guthrie: Kirchhoff, G. (1860). On the relation between the radiating and the absorbing powers of different bodies for light and heat. *Philosophical Magazine. Series 4*. 20:1–21.

Klein, L. A. and Swift, C. T. 1977. An improved model for the dielectric constant for sea water at microwave frequencies. *IEEE Transactions on Antennas and Propagation*. 25(1):104–111.

Kosolapov, V. S. and Raizer, V. Yu. 1991. Satellite microwave radiometry of the rain intensity and cloud water content (from modeling results). *Soviet Journal of Remote Sensing*. 8(5):860–878 (translated from Russian).

Kravtsov, Yu. A., Mirovskaya, Ye. A., Popov, A. Ye., Troitskiy, I. A., and Etkin, V. S. 1978. Critical effects in the thermal radiation of a periodically uneven water surface. *Izvestiya, Atmospheric and Oceanic Physics*. 14(7):522–526 (translated from Russian).

Kremer, F. and Schönhals, A. 2003. *Broadband Dielectric Spectroscopy*. Springer-Verlag, Berlin Heidelberg.

Krotikov, V. D., Mordvinkin, I. N., Pelyushenko, A. S., Pelyushenko, S. A., and Rakut, I. V. 2002. Radiometric methods of remote sensing of oil spills on water surfaces. *Radiophysics and Quantum Electronics*. 45(3):220–229 (translated from Russian).

Kutuza, B. G. 2003. Spatial and temporal fluctuations of atmospheric microwave emission. *Radio Science*. 38(3)8047:12-1–12-7.

Kuz'min, A. V. and Raizer, V. Yu. 1991. Application of the theory of excursions of a random field to the analysis of radiation from a rough surface in the quasistatic approximation. *Radiophysics and Quantum Electronics*. 34(2):128–135 (translated from Russian).

Lahtinen, J., Gasiewski, A. J., Klein, M., and Corbella, I. 2003a. A calibration method for fully polarimetric microwave radiometers. *IEEE Transactions on Geoscience and Remote Sensing*. 41(3):588–602.

Lahtinen, J., Pihlflyckt, J., Mononen, I., Tauriainen, S. J., Kemppinen, M., and Hallikainen, M. T. 2003b. Fully polarimetric microwave radiometer for remote sensing. *IEEE Transactions on Geoscience and Remote Sensing*. 41(8):1869–1878.

Landau, L. D. and Lifshitz, E. M. 1984. *Electrodynamics of Continuous Media*, 2nd edition (with L. P. Pitaevskii). Pergamon Press, Oxford, New York.

Lang, R. H., Utku, C., and Le Vine, D. M. 2003. Measurement of the dielectric constant of seawater at L-band. In *Proceedings of International Geoscience and Remote Sensing Symposium*, July 21–25, 2003, Toulouse, France, Vol. 1, pp. 19–21.

Lang, R., Zhou, Y., Utku, C., and Le Vine, D. 2016. Accurate measurements of the dielectric constant of seawater at L band. *AGU Publication Radio Science*. 51. Doi: 10.1002/2015RS00577.

Laursen, B. and Skou, N. 2001. Wind direction over the ocean determined by an airborne, imaging, polarimetric radiometer system. *IEEE Transactions on Geoscience and Remote Sensing*. 39(7):1547–1555.

Lavender, S. and Lavender, A. 2015. *Practical Handbook of Remote Sensing*. CRC Press, Boca Raton, FL.

Levin, M. L. and Rytov, S. M. 1973. "Kirchhoff" form of fluctuation-dissipation theorem for distributed systems. *Journal of Experimental and Theoretical Physics (ZhETF)*. 65:1382–1391 (translated from Russian). Internet http://www.jetp.ac.ru/cgi-bin/dn/e_038_04_0688.pdf

Le Vine, D. M., Lagerloef, G. S. E., Colomb, F. R., Yueh, S. H., and Pellerano, F. A. 2007. Aquarius: An instrument to monitor sea surface salinity from space. *IEEE Transactions on Geoscience and Remote Sensing*. 45(7):2040–2050.

Le Vine, D. M. and Utku, C. 2009. Comment on modified Stokes parameters. *IEEE Transactions on Geoscience and Remote Sensing*. 47(8):2707–2713.

Liebe, H., Hufford, G., and Takeshi, M. 1991. A model for the complex permittivity of water at frequencies below 1 THz. *International Journal of Infrared and Millimeter Waves*. 12(7):659–675.

Lin, Z., Zhang, X., and Fang, G. 2009. Theoretical model of electromagnetic scattering from 3D multi-layer dielectric media with slightly rough surfaces. In *Progress in Electromagnetics Research, PIER 96*, Vol. 96, pp. 37–62.

Liu, Q. and Weng, F. 2003. Retrieval of sea surface wind vector from simulated satellite microwave polarimetric measurements. *AGU Journal Radio Science*. 38(4):8078. Doi: 10.1029/2002RS002729.

Liu, Q., Weng, F., and English, S. 2011. An improved fast microwave water emissivity model. *IEEE Transactions on Geoscience and Remote Sensing*. 49(4):1238–1250.

Liu, W. T., Tang W., and Wentz, F. J. 1992. Perceptible water and surface humanity over the global oceans from Special Sensor Microwave Imager and European Center for Medium Range Weather Forecasts. *Journal of Geophysical Research*. 97(C2):2251–2264.

Lodge, A. E. 1989. *The Remote Sensing of Oil Slicks ((IP) Proceedings of the Institute of Petroleum)*. Wiley-Blackwell, Hoboken, New Jersey.

Martin, S. 2014. *An Introduction in Ocean Remote Sensing*, 2nd edition. Cambridge University Press, Cambridge.

Matveev, D. T. 1971. On the spectrum of microwave emission of ruffled sea surface. *Izvestiya, Atmospheric and Oceanic Physics*, 7(10):1070–1076 (in Russian).

Matzler, C. 2006. *Thermal Microwave Radiation: Applications for Remote Sensing*. The Institution of Engineering and Technology, London, UK.

Meissner, T. and Wentz, F. J. 2004. The complex dielectric constant of pure and sea water from microwave satellite observations. *IEEE Transactions on Geoscience and Remote Sensing*. 42(9):1836–1849.

Meissner, T. and Wentz, F. J. 2012. The emissivity of the ocean surface between 6 and 90 GHz over a large range of wind speeds and earth incidence angles. *IEEE Transactions on Geoscience and Remote Sensing*. 50(8):3004–3026.

Militskii, Yu. A., Rayzer, V. Yu., Sharkov, E. A., and Etkin, V. S. 1977. Scattering of microwave radiation by foamy structures. *Radio Engineering and Electronic Physics*. 22(11):46–50 (translated from Russian).

Militskii, Y. A., Rayzer, V. Y., Sharkov, E. A., and Etkin, V. S. 1978. On thermal emission of foamy structures. *Journal Technical Physics*. 48(5):1031–1033 (translated from Russian).

Mitnik, L. M. and Mitnik, M. L. 2003. Retrieval of atmospheric and ocean surface parameters from ADEOS-II Advanced Microwave Scanning Radiometer (AMSR) data: Comparison of errors of global and regional algorithms. *Radio Science*. 38(4):8065 Mar30-1–Mar30-10.

Monique, F. M. A. A., Schaap, M., de Leeuw, G., and Builtjes, P. J. H. 2010. Progress in the determination of the sea spray source function using satellite data. *Journal of Integrative Environmental Sciences.* 7(S1):159–166. Doi: 10.1080/19438151003621466.

Njoku, E. G. (ed.) 2014. *Encyclopedia of Remote Sensing (Encyclopedia of Earth Sciences Series).* Springer, New York.

Nordberg, W., Conaway, J., and Thaddeus, P. 1969. Microwave observations of sea state from aircraft. *Quarterly Journal of the Royal Meteorological Society.* 95(404):408–413. Doi: 10.1002/qj.49709540414.

Nörtemann, K., Hilland, J., and Kaatze, U. 1997. Dielectric properties of aqueous NaCl solutions at microwave frequencies. *Journal of Physical Chemistry. A.* 101(37):6864–6869. Doi: 10.1021/jp971623a.

Nosko, V. P. 1980. On the definition of the number of excursions of a random field above a fixed level. *Theory of Probability and Its Applications.* 24(3):598–602.

Odelevskiy, V. N. 1951. Calculations of the general conductivity of heterogeneous layers. *Journal of Technical Physics.* 21(6):667–685 (in Russian).

Padmanabhan, S. and Reising, S. C. 2003. Radiometric measurements of the microwave emissivity of reproducible breaking waves. In *Proceedings of International Geoscience and Remote Sensing Symposium,* July 21–25, 2000, Toulouse, France, Vol. 1, pp. 339–341. Doi: 10.1109/IGARSS.2003.1293769.

Padmanabhan, S., Reising, S. C., Asher, W. E., Raizer, V., and Gaiser, P. W. 2007. Comparison of modeled and observed microwave emissivities of water surfaces in the presence of breaking waves and foam. In *Proceedings of International Geoscience and Remote Sensing Symposium,* July 23–27, 2007, Barcelona, Spain, pp. 42–45. Doi: 10.1109/IGARSS.2007.4422725.

Pandey, P. C. and Kakar, R. K. 1982. An empirical microwave emissivity model for a foam-covered sea. *IEEE Journal of Oceanic Engineering.* 7(3):135–140.

Peake, W. H. 1959. Interaction of electromagnetic waves with some natural surfaces. *IRE Transactions on Antennas and Propagation.* 7(5):324–329. Doi: 10.1109/TAP.1959.1144736.

Petit, R. (ed.) 1980. *Electromagnetic Theory of Gratings (Topics in Current Physics).* Springer-Verlag, Berlin, Heidelberg.

Piepmeier, J. R. and Gasiewski, A. J. 2001. High-resolution passive polarimetric microwave mapping of ocean surface wind vector fields. *IEEE Transactions on Geoscience and Remote Sensing.* 39(3):606–622.

Piepmeier, J. R., Long, D. G., and Njoku, E. G. 2008. Stokes antenna temperatures. *IEEE Transactions on Geoscience and Remote Sensing.* 46(2):516–527.

Planck, M. 1914. *The Theory of Heat Radiation.* P. Blakiston's Son & Co, Philadelphia, PA.

Pospelov, M. N. 1996. Surface wind speed retrieval using passive microwave polarimetry: The dependence on atmosphere stability. *IEEE Transactions on Geoscience and Remote Sensing.* 34(5):1166–1171.

Pospelov, M. N. 2004. Wind direction signal in polarized microwave emission of sea surface under various incidence angles. *Gayana (Concepción).* 68(2):493–498. Internet http://dx.doi.org/10.4067/S0717-65382004000300032.

Potter, H., Smith, G. B., Snow, C. M., Dowgiallo, D. J., Bobak, J. P., and Anguelova, M. D. 2015. Whitecap lifetime stages from infrared imagery with implications for microwave radiometric measurements of whitecap fraction. *Journal of Geophysical Research. Oceans.* 120(11):7521–7537. Doi: 10.1002/2015JC011276.

Raicu, V. and Feldman, Y. 2015. *Dielectric Relaxation in Biological Systems: Physical Principles, Methods, and Applications.* Oxford University Press, Oxford, UK.

Raizer, V. Yu. 1992. Two phase ocean surface structures and microwave remote sensing. In *Proceedings of International Geoscience and Remote Sensing Symposium*, May 26–29, 1992, Houston, TX, Vol. 3, pp. 1460–1462. Doi: 10.1109/IGARSS.1992.578483.

Raizer, V. 2001. Modeling of sea-roughness radiometric effects for the retrieval of surface salinity at 14.3 GHz. In *Proceedings of International Geoscience and Remote Sensing Symposium*, July 9–13, 2001, Sydney, Australia, Vol. 4, pp. 1761–1763. Doi: 10.1109/IGARSS.2001.977063.

Raizer, V. 2004. Passive microwave detection of bubble wakes. In *Proceedings of International Geoscience and Remote Sensing Symposium*, September 20–24, 2004, Anchorage, Alaska, Vol. 5, pp. 3592–3594. Doi: 10.1109/IGARSS.2004.1370488.

Raizer, V. 2005. A combined foam-spray model for ocean microwave radiometry. In *Proceedings of International Geoscience and Remote Sensing Symposium*, 25–29, July 2005, Seoul, Korea, Vol. 7, pp. 4749–4752. Doi: 10.1109/IGARSS.2005.1526733.

Raizer, V. 2006. Macroscopic foam-spray models for ocean microwave radiometry. In *Proceedings of International Geoscience and Remote Sensing Symposium*, July 31–August 4, 2006, Denver, Colorado, pp. 3672–3675. Doi: 10.1109/IGARSS.2006.941.

Raizer, V. 2007. Macroscopic foam-spray models for ocean microwave radiometry. *IEEE Transactions on Geoscience and Remote Sensing*. 45(10):3138–3144. Doi: 10.1109/TGRS.2007.895981.

Raizer, V. 2008. Modeling of L-band foam emissivity and impact on surface salinity retrieval. In *Proceedings of International Geoscience and Remote Sensing Symposium*, July 6–11, 2008, Boston, MA, Vol. 4, pp. IV-930–IV-933. Doi: 10.1109/IGARSS.2008.4779876.

Raizer, V. 2009. Modeling L-band emissivity of a wind-driven sea surface. In *Proceedings of International Geoscience and Remote Sensing Symposium*, July 12–17, 2009, Cape Town, South Africa, pp. III-745–III-748. Doi: 10.1109/IGARSS.2009.5417872.

Raizer, V. 2012. Microwave scattering model of sea foam. In *Proceedings of International Geoscience and Remote Sensing Symposium*, July 22–27, 2012, Munich, Germany, pp. 5836–5839. Doi: 10.1109/IGARSS.2012.6352282.

Raizer, V. 2013. Radar backscattering from sea foam and spray. In *Proceedings of International Geoscience and Remote Sensing Symposium*, July 21–26, 2013, Melbourne, Australia, pp. 4054–4057. Doi: 10.1109/IGARSS.2013.6723723.

Raizer, V. 2014. Impedance model of sea surface at S-L-band. In *Proceedings of International Geoscience and Remote Sensing Symposium*, 13-18 July 2014, Quebec City, Quebec, Canada, pp. 4400–4403. Doi: 10.1109/IGARSS.2014.6947466.

Raizer, V. Yu. and Cherny, I. V. 1994. Microwave diagnostics of ocean surface. "Mikrovolnovaia diagnostika poverkhnostnogo sloia okeana." Gidrometeoizdat. Sankt-Peterburg. Library of Congress, LC classification (full) GC211.2 .R35 1994. (in Russian).

Raizer, V. Yu. and Sharkov, E. A. 1981. Electrodynamic description of densely packed dispersed system. *Radiophysics and Quantum Electronics*. 24(7):553–557 (translated from Russian).

Raizer, V. Yu., Zaitseva, I. G., Aniskovich, V. M., and Etkin, V. S. 1986. Determining sea ice physical parameters from remotely sensed microwave data in the 0.3–18 cm. *Soviet Journal of Remote Sensing*. 5(1):29–42 (translated from Russian).

Ray, P. S. 1972. Broadband complex refractive indices of ice and water. *Applied Optics*. 11(8):1836–1844.

Rayzer, V. Yu., Sharkov, E. A., and Etkin, V. S. 1975. Influence of temperature and salinity on the radio emission of a smooth ocean surface in the decimeter and meter bands. *Izvestiya, Atmospheric and Oceanic Physics*. 11(6):652–655 (translated from Russian).

Robinson, I. S. 2010. *Discovering the Ocean from Space: The Unique Applications of Satellite Oceanography*. Springer, Berlin, Heidelberg.

Robitaille, P.-M. 2009. Kirchhoff's Law of Thermal Emission: 150 Years. *Progress in Physics*. 4:3–13.

Rose, L. A., Asher, W. E., Reising, S. C., Gaiser, P. W., Germain, K. M. St., Dowgiallo, D. J., Horgan, K. A., Farquharson, G., and Knapp, E. J. 2002. Radiometric measurements of the microwave emissivity of foam. *IEEE Transactions on Geoscience and Remote Sensing*. 40(12):2619–2625.

Rozenkranz, P. V. and Staelin, D. H. 1972. Microwave emissivity of ocean foam and its effect on nadiral radiometric measurements. *Journal of Geophysical Research*. 77(33):6528–6538.

Ruf, C. S. 1998. Constraints on the polarization purity of a Stokes microwave radiometer. *Radio Science*. 33(6):1617–1639.

Rytov, S. M., Kravtsov, Yu. A., and Tatarskii, V. I. 1989. *Principles of Statistical Radiophysics*. Vol. 3. Springer-Verlag, Berlin.

Sadovsky, I. N., Kuzmin, A. V., and Pospelov, M. N. 2009. Dynamics of short sea wave spectrum estimated from microwave radiometric measurements. *IEEE Transactions on Geoscience and Remote Sensing*. 47(9):3051–3056.

Salisbury, D. J., Anguelova, M. D., and Brooks, I. M. 2014. Global distribution and seasonal dependence of satellite-based whitecap fraction. *Geophysical Research Letters*. 41(5):1616–1623. Doi: 10.1002/2014GL059246.

Sasaki, Y., Asanuma, I., Muneyama, K., Naito, G., and Suzuki, T. 1987. The dependence of sea-surface microwave emission on wind speed, frequency, incident angle and polarization over the frequencies range from 1 to 40 GHz. *IEEE Transactions on Geoscience and Remote Sensing*. 25(2):138–146.

Savelyev, I. B., Anguelova, M. D., Frick, G. M., Dowgiallo, D. J., Hwang, P. A., Caffrey, P. F., and Bobak, J. P. 2014. On direct passive microwave remote sensing of sea spray aerosol production. *Atmospheric Chemistry and Physics*. 14:11611–11631. Doi: 10.5194/acp-14-11611-2014.

Scou, N. 1989. *Microwave Radiometer Systems. Design & Analysis*. Artech House, Norwood, MA.

Sharkov, E. A. 2003. *Passive Microwave Remote Sensing of the Earth: Physical Foundations*. Springer Praxis Books, Chichester, UK.

Sharkov, E. A. 2007. *Breaking Ocean Waves: Geometry, Structure and Remote Sensing*. Praxis Publishing, Chichester, UK.

Shestopaloff, Yu. K. 2011. Polarization invariants and retrieval of surface parameters using polarization measurements in remote sensing applications. *Applied Optics*. 50(36):6606–6616.

Shutko, A. M. 1985. The status of the passive microwave sensing of the waters—Lakes, seas, and oceans—Under the variation of their state, temperature, and mineralization (salinity): Models, experiments, examples of application. *IEEE Journal of Oceanic Engineering*. 10(4):418–437. Doi: 10.1109/JOE.1985.1145121.

Shutko, A. M. 1986. *Microwave Radiometry of A Water Surface and The Ground*. Nauka (in Russian), Moscow.

Sihvola, A. 1999. *Electromagnetic Mixing Formulas and Applications, IEE Electromagnetic Waves Series*, Vol. 47. The Institute of Electrical Engineers.

Skou, N. 1986. Microwave radiometry for oil pollution monitoring, measurements, and systems. *IEEE Transactions on Geoscience and Remote Sensing.* 24(3):360–367.

Skou, N. and Laursen, B. 1998. Measurement of ocean wind vector by an airborne, imaging, polarimetric radiometer. *Radio Science.* 33(3):669–675.

Smith, P. M. 1988. The emissivity of sea foam at 19 and 37 GHz. *IEEE Transactions on Geoscience and Remote Sensing.* 29(5):541–547.

Somaraju, R. and Trumpf, J. 2006. Frequency, temperature and salinity variation of the permittivity of seawater. *IEEE Transactions on Antennas and Propagation.* 54(11):3441–3448.

Stogryn, A. 1971. Equations for calculating the dielectric constant for saline water. *IEEE Transactions on Microwave Theory and Technology.* 19(8):733–736.

Stogryn, A. 1972. The emissivity of sea foam at microwave frequencies. *Journal of Geophysical Research.* 77(9):1698–1666.

Stratton, J. A. 1941. *Electromagnetic Theory.* McGraw-Hill Book Company.

Swift, C. T. 1974. Microwave radiometric measurements of the Cape God Canal. *Radio Science.* 9(7):641–653.

Swift, C. T. and MacIntosh, R. E. 1983. Considerations for microwave remote sensing of ocean-surface salinity. *IEEE Transactions on Geoscience and Remote Sensing.* 21(4):480–491.

Tang, C. C. H. 1974. The effect of droplets in the air-sea transition zone on the sea brightness temperature. *Journal of Physical Oceanography.* 4:579–593.

Tatarskii, V. I. 1961. *Wave Propagation in a Turbulent Medium.* Translated by A. Silverman. McGraw-Hill, New York.

Tatarskii, V. V. and Tatarskii, V. I. 1996. Non-Gaussian statistical model of the ocean surface for wave-scattering theories. *Waves in Random Media.* 6(4):419–435.

Tikhonov, A. N. and Arsenin, V. Y. 1977. *Solutions of Ill-Posed Problems.* V.H. Winston.

Tinga, W. R., Voss, W. A. G., and Blossey, D. F. 1973. Generalized approach to multi-phase dielectric mixture theory. *Journal of Applied Physics.* 44(9):3897–3902.

Trokhimovski, Yu. G. 2000. Gravity–capillary wave curvature spectrum and mean-square slope retrieved from microwave radiometric measurements (Coastal Ocean Probing Experiment). *Journal of Atmospheric and Oceanic Technology.* 17(9): 1259–1270. http://dx.doi.org/10.1175/1520-0426(2000)017<1259:GNCWCS>2.0.CO;2

Trokhimovski, Yu. G., Bolotnikova, G. A., Etkin, V. S., Grechko, S. I., and Kuzmin, A. V. 1995. The dependence of S-band sea surface brightness temperature on wind vector at normal incidence. *IEEE Transactions on Geoscience and Remote Sensing.* 33(4):1085–1088.

Trokhimovski, Y. G., Irisov, V. G., Westwater, E. R., Fedor, L. S., and Leuski, V. E. 2000. Microwave polarimetric measurements of the sea surface brightness temperature from a blimp during the Coastal Ocean Probing Experiment (COPE). *Journal of Geophysical Research.* 105(C3):6501–6516.

Trokhimovski, Y. G., Kuzmin, A. V., Pospelov, M. N., Irisov, V. G., and Sadovsky, I. N. 2003. Laboratory polarimetric measurements of microwave emission from capillary waves. *Radio Science.* 38(3):MAR 4-1–MAR 4-7.

Tsang, L., Chen, C. T., Chang, A. T. C., Guo, J., and Ding, K. H. 2000a. Dense media relative transfer theory based on quasi-crystalline approximation with applications to passive microwave remote sensing of snow. *Radio Science.* 35(3):731–749.

Tsang, L., Kong, J. A., and Ding, K. H. 2000b. Scattering and emission by layered media. In *Scattering of Electromagnetic Waves: Theories and Application.* J. A. Kong (ed.), pp. 203–207. John Wiley and Sons, New York, NY.

Tsang, L., Kong, J. A., and Shin, R. T. 1985. *Theory of Microwave Remote Sensing*. Wiley-Interscience, New York.

Tsang, L., Njoku, E., and Kong, J. A. 1975. Microwave thermal emission from a stratified medium with nonuniform temperature distribution. *Journal of Applied Physics*. 46(12):5127–5133.

Ulaby, F. T. and Long, D. G. 2013. *Microwave Radar and Radiometric Remote Sensing*. University of Michigan Press, Ann Arbor, Michigan.

Ulaby, F. T., Moore, R. K., and Fung, A. K. 1981,1982,1986. *Microwave Remote Sensing. Active and Passive* (in three volumes), Advanced Book Program, Reading and Artech House, MA.

van de Hulst, H. 1957. *Light-Scattering by Small Particles*. Wiley, New York; also Dover, New York, 1981.

Van Melle, M. J., Wang, H. H., and Hall, W. F. 1973. Microwave radiometric observations of simulated sea surface conditions. *Journal of Geophysical Research*. 78(6):969–976. Doi: 10.1029/JC078i006p00969.

Veron, F., Melville, W. K., and Lenain L. 2009. Measurements of ocean surface turbulence and wave–turbulence interactions. *Journal of Physical Oceanography*. 39(9):2310–2323. http://dx.doi.org/10.1175/2009JPO4019.1.

Von Hippel, A. R. 1995. *Dielectrics and Waves*, 2nd edition. Artech House, Boston.

Voronovich, A. 1994. Small-slope approximation for electromagnetic wave scattering at a rough interface of two dielectric half-space. *Waves in Random Media*. 4(3):337–367.

Voronovich, A. 1996. On the theory of electromagnetic waves scattering from the sea surface at low grazing angles. *Radio Science*. 31(6):1519–1530.

Voronovich, A. 1999. *Wave Scattering from Rough Surfaces (Springer Series on Wave Phenomena)*, 2nd edition. Springer, Berlin, Heidelberg.

Vorsin, N. N., Glotov, A. A., Mirovskiy, V. G., Raizer, V. Yu., Troitskii, I. A., Sharkov, E. A., and Etkin, V. S. 1984. Direct radiometric measurements of sea foam. *Soviet Journal of Remote Sensing*. 2(3):520–525 (translated from Russian).

Webster, W. J., Wilheit, T. T., Ross, D. B., and Gloersen, P. G. 1976. Spectral characteristics of the microwave emission from a wind-driven foam-covered sea. *Journal of Geophysical Research*. 81(18):3095–3099.

Wei, E. B. 2011. Microwave vector radiative transfer equation of a sea foam layer by the second-order Rayleigh approximation. *Radio Science*. 46(5):RS5012–RS5013.

Wei, E. B. 2013. Effective medium approximation model of sea foam layer microwave emissivity of a vertical profile. *International Journal of Remote Sensing*. 34(4):1180–1193.

Wei, E. B. Liu, S. B. Wang, Z. Z. Liu, J. Y., and Dong, S. 2014a. Emissivity measurements and theoretical model of foam-covered sea surface at C-band. *International Journal of Remote Sensing*. 35(4):1511–1525.

Wei, E.-B., Liu, S.-B., Wang, Z.-Z., Tong, X.-L., Dong, S., Li, B., and Liu, J.-Y. 2014b. Emissivity measurements of foam-covered water surface at L-band for low water temperatures. *Remote Sensing*. 6(11):10913–10930. Doi: 10.3390/rs61110913.

Wentz, F. J. 1975. A two-scale scattering model for foam-free sea microwave brightness temperatures. *Journal of Geophysical Research*. 80(24):3441–3446.

Wentz, F. J. 1992. Measurement of oceanic wind vector using satellite microwave radiometers. *IEEE Transactions on Geoscience and Remote Sensing*. 30(5):960–972.

Wentz, F. J. 1997. A well-calibrated ocean algorithm for SSM/I. *Journal of Geophysical Research*. 102(C4):8703–8718.

Wilheit, T. T. 1979. A model for the microwave emissivity of the ocean's surface as a function of wind speed. *IEEE Transactions on Geoscience Electronics.* 17(4):244–249.

Williams, G. 1971. Microwave emissivity measurements of bubble and foam. *IEEE Transactions on Geoscience Electronics.* 9(4):221–224.

Wilson, W. J., Yueh, S. H., Dinardo, S. J., Chazanoff, S. F., Kitiyakara, A., Li, F., and Rahmat-Samii, Y. 2001. Passive active L- and S-band (PALS) microwave sensor for ocean salinity and soil moisture measurements. *IEEE Transactions on Geoscience and Remote Sensing.* 39(5):1039–1048.

Woodhouse, I. H. 2005. *Introduction to Microwave Remote Sensing.* CRC Press, Boca Raton, FL.

Wu, S. T. and Fung, A. K. 1972. A noncoherent model for microwave emissions and backscattering from the sea surface. *Journal of Geophysical Research.* 77(30):5917–5929. Doi: 10.1029/JC077i030p05917.

Yueh, S. H. 1997. Modelling of wind direction signals in polarimetric sea surface brightness temperatures. *IEEE Transactions on Geoscience and Remote Sensing.* 35(6):1400–1418.

Yueh, S. and Chaubell, J. 2012. Sea surface salinity and wind retrieval using combined passive and active L-band microwave observations. *IEEE Transactions on Geoscience and Remote Sensing.* 50(4):1022–1032.

Yueh, S., Dinardo, S., Fore, A., and Li, F. 2010. Passive and active L-band microwave observations and modeling of ocean surface winds. *IEEE Transactions on Geoscience and Remote Sensing.* 48(8):3087–3100.

Yueh, S. H., Kwok, R., Li, F. K., Nghiem, S. V., Wilson, W. J., and Kong, J. A. 1994a. Polarimetric passive remote sensing of ocean wind vector. *Radio Science.* 29(4):799–814.

Yueh, S. H., Nghiem, S. V., Kwok, R., Wilson, W. J., Li, F. K., Johnson, J. T., and Kong, J. A. 1994b. Polarimetric thermal emission from periodic water surfaces. *Radio Science.* 29(1):87–96.

Yueh, S. H., Tang, W., Fore, A., Neumann, G., Hayashi, A., Freedman, A., Chaubell, J., and Lagerloef, G. 2013. L-band passive and active microwave geophysical model functions of ocean surface winds and applications to Aquarius retrieval. *IEEE Transactions on Geoscience and Remote Sensing.* 51(9):4619–4632.

Yueh, S. H., Wilson, W., Dinardo, S., and Hsiao, S. V. 2006. Polarimetric microwave wind radiometer model function and retrieval testing for WindSat. *IEEE Transactions on Geoscience and Remote Sensing.* 44(2):584–596.

Yueh, S. H., Wilson, W. J., Dinardo, S. J., and Li, F. K. 1999. Polarimetric microwave brightness signatures of ocean wind directions. *IEEE Transactions on Geoscience and Remote Sensing.* 37(2):949–959.

Yueh, S. H., Wilson, W. J., Li, F. K., Nghiem, S. V., and Ricketts, W. B. 1995. Polarimetric measurements of sea surface brightness temperature using an aircraft K-band radiometer. *IEEE Transactions on Geoscience and Remote Sensing.* 33(1):85–92.

Yueh, S. H., Wilson, W. J., Li, F. K., Nghiem, S. V., and Ricketts, W. B. 1997. Polarimetric brightness temperatures of sea surfaces measured with aircraft K- and Ka-band radiometers. *IEEE Transactions on Geoscience and Remote Sensing.* 35(5):1177–1187.

4

Simulation and Prediction of Ocean Data

A novel combined digital framework that is dedicated to the analysis and prediction of complex microwave remotely sensed data (signals, images, signatures) is presented in this chapter. The framework operates with multifactor microwave stochastic models and includes elements of digital signal/image processing and computer vision. These algorithms are described in the literature (Pratt 2001; Chen 2007; Szeliski 2011; Lillesand et al. 2015).

The framework provides digital modeling and simulations of a variety of ocean microwave remote sensing data, scenes, and scenarios. Numerical examples and selected results are presented and discussed in order to demonstrate microwave capability assessment. The framework is designed as a flexible computer tool for scientific research but not for operational purposes (Raizer 1998, 2002, 2005, 2011).

4.1 Basic Description

A framework represents a generalized forward linear multifactor model, describing statistical combinations of the microwave emission contributions induced by different environmental factors. A framework implementation provides a spatial averaging and spatial intermittent connectivity of the contributions at variable observation conditions.

Figure 4.1 shows a framework flowchart. It includes several consecutive operations combined into a unit algorithm. They are the following.

The microwave emission contributions related to individual environmental factors are defined using the corresponding electromagnetic models (Chapter 3). A statistical ensemble of these individual contributions forms a composite microwave radio-brightness scene, which is characterized by a spatial probability distribution function (pdf). As a result, an *observation multifactor composite remote sensing microwave model (RSMM)* is created. An observation process is invoked through the convolution between actual (modeled) radio-brightness scene and the point spreading function (psf) providing averaging and filtering. This framework allows us to investigate different microwave radiometric data and convert them into a computer vision product.

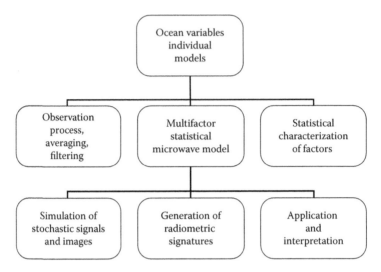

FIGURE 4.1
A framework flowchart for modeling and simulation of microwave data.

4.2 Multifactor Microwave Model

Multifactor models employ multiple factors to predict and explain the behavior of composite phenomena and/or arbitrary systems. Historically, mathematical theory and multifactor models are developed and used for the analysis of financial markets, but we believe that this economic concept can also be applied in remote sensing. Indeed, a large number of random variables should be taken into account in order to estimate a total electromagnetic response with some probability. The analogy is understandable. Important geophysical examples are low forecasting hazardous events and dynamic complex ocean scenes that can be investigated using remote sensing methods.

Multifactor spectral microwave model of the ocean surface can be represented in terms of the total brightness temperature contrast of a given geophysical scene as the following:

$$\Delta T_{B\lambda}(\vec{r}, t) = \sum_{i=1}^{N}\left[\sum_{k=1}^{M}\frac{\partial T_{Bi}\left(\vec{r}, t; q_i^k\right)}{\partial q_i^k}\Delta q_i^k W_i(\vec{r}, t) + \Psi_i^k\right], \qquad (4.1)$$

where ΔT_{Bi} is the brightness temperature variation induced by factors, Δq_i^k is the variation of the parameters, W_i is the statistical weight coefficient, Ψ_i^k is the corresponding error term, N is the number of participation factors, M is the number of input parameters related to each factor, $\vec{r} = \{x, y\}$ is the coordinate vector, and t is time. This model operates by deterministic and/or statistical parameters and distributions that are related to individual hydro-physical factors.

For example, in the simple practical case of stationary statistically isotropic wind-generated ocean surface involving three participating factors—roughness, foam, and whitecap—the microwave model can be written as

$$\Delta T_{B\lambda}(V) = \sum_{i=1}^{3} \Delta T_{Bi} W_i(V), \quad \text{or} \quad \Delta T_{B\lambda} = \Delta T_{B1} W_1 + \Delta T_{B2} W_2 + \Delta T_{B3} W_3, \quad (4.2)$$

$$W_1 + W_2 + W_3 = 1, \quad W_1 + W_2 = W_f,$$

$$W_1 = \frac{1}{1+R_f} W_f, \quad W_2 = \frac{R_f}{1+R_f} W_f,$$

$$W_f = aV^b, \quad R_f = A + BV,$$

$$\Delta T_{B1,2,3} = \frac{\partial T_{B1,2,3}(T,S)}{\partial T} \Delta T + \frac{\partial T_{B1,2,3}(T,S)}{\partial S} \Delta S,$$

where $\Delta T_{B1,2,3}$ are brightness temperature contrasts induced by individual factors (calculated using electromagnetic models, Chapter 3), V is wind speed, T is sea surface temperature, S is salinity, W_3, W_2, W_3 are area fractions corresponding to surface roughness, foam streaks, and whitecap, W_f is the total foam+whitecap area fraction, R_f is the ratio of foam-to-whitecap area fractions, and a, b, A, B are empirical constants.

The model (4.2) provides estimations of radiation-wind dependency of the total contrast $\Delta T_{B\lambda}(V)$ at a specified wavelength λ in the presence of three factors—surface roughness, foam streaks, and whitecap. Simultaneously, variations of two physical parameters—sea surface temperature and salinity—are formally taken into account. This particular model can be expanded on more complicated cases when others geophysical parameters are involved as well.

4.3 Mathematical Formulation

A mathematical formulation of the generalized forward linear multifactor statistical RSMM can be written in terms of the brightness temperature matrix scene $\overline{\overline{T}}_{A,BS}$ using the following two separated parts:

$$\overline{\overline{T}}_{BS} = \sum_{i=1}^{N} \overline{\overline{T}}_{Bi} \otimes \overline{\overline{H}}_{Pi}, \quad \text{Stochastic scene} \quad (4.3)$$

$$\bar{\bar{T}}_A = \bar{P} * * \bar{\bar{T}}_{BS} + \bar{\eta} + \bar{\mu}, \quad \text{Observation process} \tag{4.4}$$

where $\bar{\bar{T}}_A$ is the desired (measured) scene, $\bar{\bar{T}}_{BS}$ is the actual (modeled) stochastic multifactor scene, $\bar{\bar{T}}_{Bi}$ is the actual deterministic scene for the i-factor, $\bar{\bar{T}}_{Bi} \otimes \bar{\bar{H}}_{pi}$ is the Kronecker product of two matrices, $\bar{\bar{H}}_{Pi}$ is the stochastic weighing matrix describing a random field for the i-factor and operated with the corresponding pdf, $\bar{P}$ is the vector psf, $\bar{\eta}$ is additive instrument noise, $\bar{\mu}$ is unobserved (hidden) geophysical noise, N is the number of participating geophysical factors.

In Equation 4.3 model matrix $\bar{\bar{T}}_{Bi}$ is defined through the microwave contributions from each geophysical (or hydrodynamic) factor, whereas model matrix $\bar{\bar{T}}_{BS}$ is defined as a multifactor statistical microwave response. An observation process (4.4) describes a resulting averaged and filtered stochastic microwave scene. The vector psf $\bar{P}$ is defined in accordance with technical parameters of the instrument and geometry of observations. Instrument noise $\bar{\eta}$ is modeled by the Gaussian distribution function; geophysical noise $\bar{\mu}$ is defined as randomness process (field) describing unobservable contributions from certain geophysical variables. For example, if it is necessary to extract information concerning hydro-physical parameters from microwave data, the influences of Earth's atmosphere are considered as a geophysical noise.

Because the observation process is a function of the desired geophysical data sets, two categories of intelligent data acquisition can be considered: detection and estimation. In the theory of statistical signal processing and communication, these two categories are connected (overlapped) with each other, which allows for the detection and recognition of target variables with certain probability. Both observation errors and process noise are estimated as well.

In the case of ocean observations, especially those related to the detection of localized hydrodynamic events (but not fields of wind, temperature, or salinity), the classical acquisition schemes may not work due to a great variety of oceanic environment, noisy processes, and unpredictability of the event behavior in space and time. Moreover, there is no ultimate theoretical model allowing for the solution of the inverse remote sensing problem at nonstationary ocean conditions. Therefore, one available option is direct numerical modeling and simulation of microwave data, followed by specification and classification of microwave signatures. The practical implementation of such a complex algorithm requires multiple operations. In particular, digital utilization (4.3) and (4.4) involves a large number of input and output parameters and variables that makes it difficult to realize the overall picture in advance. In fact, a research strategy requires the implementation of computer experiments, invoking digital methods of data processing.

4.4 Examples of Data Simulations

A gallery of microwave radiometric signals and images typical for ocean environments is presented. In fact, these data are computational replicas of real-world raw experimental records, registered by sensitive passive microwave radiometers. The chosen examples show up the content and stochastic *structuralization* of ocean radiometric data. This material allows the reader to get a deeper insight into the problems of high-resolution measurements and understand better the principles of data processing and interpretations.

4.4.1 Environmental Signatures

The environmental ocean microwave background is associated, first of all, with large-scale surface dynamics. In the case of two-factor representation (surface roughness + foam), the actual microwave scene $T_B(x,y) = F\{\Delta T_s(x,y); \Delta T_f(x,y)\}$ is computed automatically through some numerical spread operator $F\{\ldots\}$. To reduce uncertainties, we apply a linear operator, which is a weighted sum of the microwave contributions from surface roughness and foam/whitecap, $F\{\ldots\} = \Delta T_s(x,y) \cdot W_s + \Delta T_f(x,y) \cdot W_f$, where $W_{s,f}$ are the corresponding area fractions (dependent on wind speed). Such an approach is suitable for observations with large, spatially temporal averaging of radiometric signals. Linear operation allows us to create realistic-enough microwave scenes and distinct macro-textures related to high wind situations and foam/whitecap activity. To reduce possible errors arising from computer syntheses, the resulting image $T_B(x,y)$ is filtered by brightness thresholds and certain spatial frequencies.

Figure 4.2 demonstrates several typical radiometric profiles and images, simulated for the ocean surface at different situations. Surface roughness yields relatively small variations of radiometric signals (~2–3 K) and foam yields large variations of signals in the form of amplitude impulses (~5–10 K). Modeling and simulations of radiometric profiles were made simultaneously at K, X, C, S, and L bands. These data demonstrate important effects: stochastic (noise) character of ocean radiometric signals and their multiband correlations. In fact, these effects occur due to the joint statistical impact of surface roughness and foam coverage. Such types of radiometric signals have been observed in field experiments (from ship and aircraft platforms) many times. As a whole, amplitude trends and fluctuations of radiometric signals reflect nonstationary surface conditions at variable wind. For example, it could be limited wind fetch. However, in the real world, ocean radiometric signals are defined not only by environmental conditions but also by an observation process (i.e., temporally spatial averaging) that is important for the specification of the relevant signatures.

Specific variations of radiometric signals (signatures) can occur in the presence of localized ocean futures and/or surface disturbances as well. Among

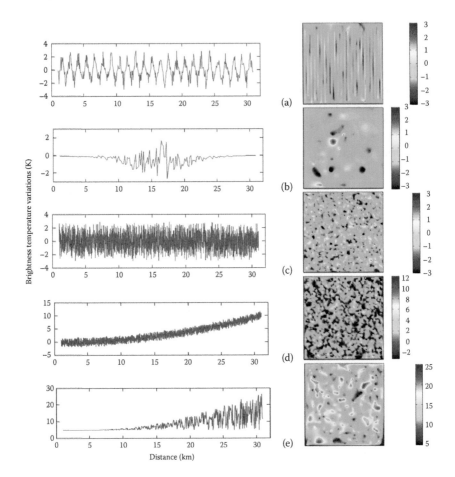

FIGURE 4.2
Microwave radiometric data simulated digitally for different ocean conditions at Ku band. Profiles (left panel): (a) periodic (internal waves); (b) anomaly (soliton); (c) normal stationary random process (uniform wind); (d) normal process with amplitude trend and constant r.m.s. deviation (nonuniform wind); (e) normal process with amplitude trend and variable r.m.s. deviation (nonuniform increased wind with wave breaking and foam/whitecap). Images (right panel). Size of scene ~30 km × 30 km. (a) and (b) deterministic models; (c–e) statistical models (stochastic ocean microwave background). (Adapted from Raizer, V. 2005. In *Proceedings of International Geoscience and Remote Sensing Symposium*, Vol. 1, pp. 268–271. Doi: 10.1109/IGARSS.2005.1526159.)

them, the most probable causes are roughness change in restricted areas due to hydrodynamic wave–wave interactions, modulations, and instabilities, and wave breaking actions (or microbreaking) as well. In these cases, it is difficult to evaluate possible microwave signatures without complementary information or model data.

In Figure 4.3, we demonstrate some typical model examples. These signatures are distinguished by several characteristics: (1) a set of high-contrast

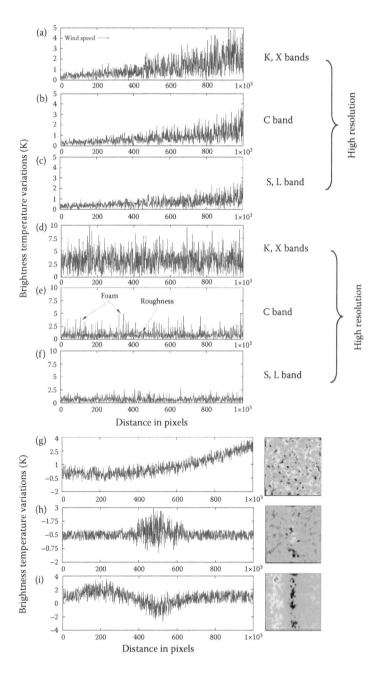

FIGURE 4.3
Computer simulations of high-resolution multiband radiometric data for variable ocean surface conditions. Digital examples. Left panel—time series signals (profiles); Right panel—images. Realizations at K, X, C, S, and L bands: (a–c) wind fetch; (d–f) impact of roughness and foam. Realizations at L band only: (g) wind action; (h) temperature front; (i) salinity anomaly.

impulse-like fluctuations is associated with wave breaking and foam/white-cap activity; (2) low-contrast periodic-like signals can be induced by modulations of surface roughness due to, for example, interactions of internal and surface waves manifested as alternating slicks and rip currents; (3) bombing-type deviations that can occur under the influence of surface wakes, thin foam patches, surface currents, or fronts; (4) extended variations of radiometric signals associated with localized sea surface temperature–salinity variations caused by thermohaline processes.

From the presented data, it follows that any combinations or random mixing of the microwave emission contributions produce complex radiometric data sets, including multicontrast radio-brightness texture images. Ocean microwave textures represent a novel class of remote sensing information; they characterize environmental processes and fields through passive microwave pictures. Synthesized complex textures provide the detection and recognition of relevant radiometric signatures, for example, related to foam/whitecap coverage. These signatures represent extended and/or localized image objects (spots) by analogy with optical data. However, experimental verification of such a model and detailed realizations have not been done yet; this remains a challenging task for high-resolution passive microwave radiometry.

4.4.2 Roughness–Salinity–Temperature Anomalies

Roughness–salinity–temperature anomalies (RSTA) represent complex *thermohydrodynamic* features associated with simultaneous variations of sea surface roughness, surface salinity, and temperature. Experimental oceanographic data show that the near-surface layer of the upper ocean, including the air–water interface and electromagnetic skin layer, is usually unstable and nonuniform. These conditions occur under the influence of many environmental processes: thermohaline (i.e., joint salinity, temperature, and density) circulations, double-diffusive and convective processes, turbulent mixing, hydrodynamic interactions, wave modulations, currents, wave breaking, and wind actions.

In this context, natural RSTA are abundant, fascinating, and important geophysical objects for remote sensing studies.

The most probable surface manifestations of RSTA can be the following: (1) a double-diffusive instability called "salt fingers"; (2) temperature fronts, thermohaline "wakes," saltwater intrusions, or freshwater injections; and (3) strong rip currents referred to as "suloy."

In this case, the total brightness temperature contrast can be calculated using the resonance model (3.12) as the sum of three parts:

$$\Delta T_B = \alpha \Delta T_{Brough} + \beta \Delta T_{Bsal} + \gamma \Delta T_{Btemp}, \tag{4.5}$$

$$\Delta T_{Brough} = B_T(k_0)\frac{\partial \Theta_T}{\partial E_k}\Delta E_k,$$

$$\Delta T_{Bsal} = B_T(k_0)\frac{\partial \Theta_T}{\partial s}\Delta s, \tag{4.6}$$

$$\Delta T_{Btemp} = B_T(k_0)\frac{\partial \Theta_T}{\partial t}\Delta t,$$

where T_{Brough}, T_{Bsal}, and T_{Btemp} are brightness temperature contrasts induced by variations of surface roughness, salinity, and temperature, respectively; ΔE_k, Δs, and Δt are variations of the wave number spectrum (describing roughness change), salinity, and temperature, respectively; $\Theta_T(E_k,k_0;t,s)$ is the brightness temperature integrant delivered from Equation 3.12; k_0 is the electromagnetic wave number; k is the surface wave number; $B_T(k_0)$ is a constant; and α, β, γ are the weight coefficients.

Sea surface physical parameters can be measured using S and L band microwave radiometers. To illustrate this well-known fact, we compute radiation-wind dependencies of emissivity (Figure 4.4) and the temperature–salinity sensitivity of the brightness temperature shown in the form of two-channel cluster diagram (Figure 4.5). These results demonstrate a possibility to

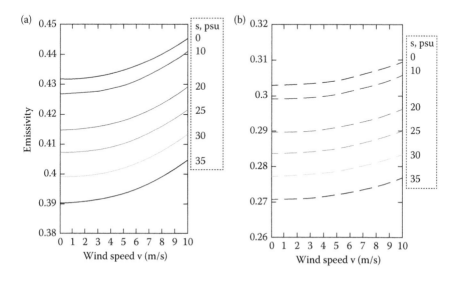

FIGURE 4.4
Radiation-wind dependencies of emissivity at L band (1.4 GHz) computed using resonance model (3.19). View angle is 37°. Salinity is varied in the range s = 0...35 psu. (a) Vertical and (b) horizontal polarizations.

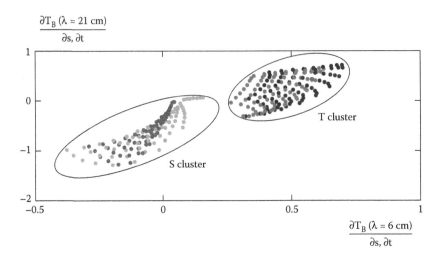

FIGURE 4.5

Two-channel T–S cluster diagram showing the sensitivity of the brightness temperature to sea surface temperature and salinity. Incidence angle is 37°, vertical polarization. Range of temperature and salinity: $t = 0$–40°C and $s = 0$–40 psu. S cluster and T cluster are well distinguished and correspond to derivatives $\partial/\partial s$ and $\partial/\partial t$ at wind speed 3...10 m/s.

distinguish effects from sea surface temperature, salinity, and roughness if applied at least two-channel low-frequency (in the range 1–3 GHz) remote sensing measurements.

Figure 4.6 illustrates a hypothetical example of RSTA signatures, generated digitally at the L band. Input data contain three components imitating the following conditions: (a) surface roughness anomaly—slick—modeled through amplitude transformations of power wave number spectrum $F(K) \sim AK^{-n}$, where A and n are parameters, (b) gradient of sea surface salinity $\vec{\nabla}s$ (salt wave), and (c) gradient of sea surface temperature $\vec{\nabla}t$ (thermal wave). Output data represent a stochastic composite radio-brightness picture (d). The appearing robust mosaic radio-brightness textures in the picture allows us to assume that oceanic RSTA are potentially detectable using high-resolution S–L band radiometry and imagery (Raizer 2010).

4.5 Composition Multiband Imagery

Another example of modeling and simulation is related to the multiband microwave imagery of the ocean surface in the presence of different environmental factors (Raizer 2011). Figure 4.7 presents two multiband sets of stochastic ocean microwave radiometric images (or digital pictures), generated simultaneously at high and low spatial resolutions (in pixels). We employ

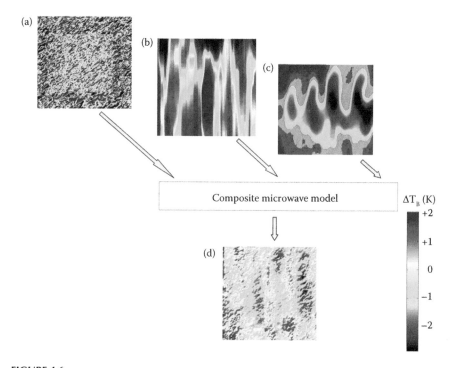

FIGURE 4.6
Roughness–salinity–temperature anomaly (RSTA). Computer simulation: (a–c) input components; (d) resulting RSTA microwave image. Color bar shows the radio-brightness contrast at L band and at 37° incidence (vertical polarization). (From Raizer 2010. In *Proceedings of International Geoscience and Remote Sensing Symposium*, pp. 3174–3177. Doi: 10.1109/IGARSS.2010.5651356.)

a four-factor statistic model with the following factors: (1) wind-generated roughness, that is, regular wave spectrum, (2) two-phase dispersed structures: foam\whitecap\spray\bubbles, (3) gradient of SST, and (4) gradient of SSS. Eventually, these factors yield different contributions to ocean emissivity.

Radio-brightness variations and contrast features (signatures) appear in the pictures as a result of multifactor band-dependent microwave emission impacts. The observability of signatures, their color, shape, and texture content are defined by the statistical distribution of microwave contributions and spatial resolution as well. Indeed, the effect of image change and "signature smoothing" due to the reduction of the pixel resolution occurs at all bands. In the case of high-resolution pictures (Figure 4.7, upper panel), there are multicontrast distinct radiometric features, whereas in the case of low-resolution pictures (Figure 4.7, lower panel), the features have mostly extended and monotonic characters. The apparent complexity of these pictures is a result of *stochastic intermixing* and/or randomization of emissivity. Such microwave effects can be observed at variable wind or nonstationary surface conditions.

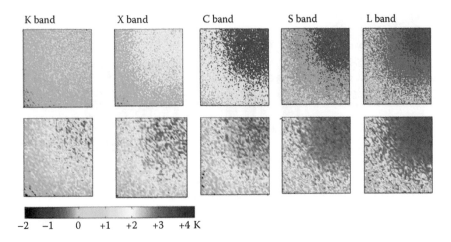

FIGURE 4.7

Two multiband sets of stochastic ocean microwave radiometric images generated simultane-ously at different pixel resolutions (four-factor model). Size of each image is 2048 × 2048 pixels. Upper panel—high resolution (grid resolution is 100 pixels) and lower panel—low resolu-tion (grid resolution is 50 pixels). Specification of bands: K band (18.7 GHz at 1.6 cm); X band (10.7 GHz at 2.8 cm); C band (6.9 GHz at 4.3 cm); S band (2.6 GHz at 8.6 cm); L band (1.4 GHz at 21 cm).

For example, in the case of fetch-limited wind-wave growth when joint geo-physical variations of SST and SSS fields yield moderate indirect contribu-tions into radiometric signals and images at S and L bands (Raizer 2009). The most valuable (measurable) image effects may occur at K, X, and C bands due to wind-generated surface roughness and breaking waves.

The presented digital examples demonstrate the following: (1) joint sta-tistical character of the microwave contributions result in the complexity and variety of multiband data; this sometimes leads to uncertainty in data analysis and interpretation and (2) an observation process (resolution, aver-aging, filtering) plays an important role in the selection and extraction of the relevant geophysical information (signatures) from ocean high-resolution imagery.

A more sophisticated digital example has been developed early in order to predict hypothetical passive microwave pictures of an ocean tsunami. An imaging model was created in accordance with a space-based microwave radiometric constellation concept (Myers 2008; Myers et al. 2008). The so-called "tsunami microwave signatures" represent periodic-type variations of the brightness temperature (~ −2 + +3 K), associated with long-period spatial-wave modulations of ocean surface roughness. A model proposed invokes an observation geometry, mapping, and timing that provide digi-tal simulations of tsunami microwave pictures at different spatial resolution (Figure 4.8).

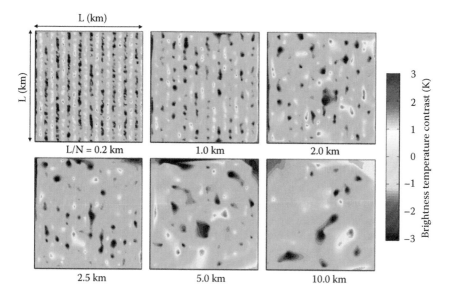

FIGURE 4.8

Passive microwave portrait of oceanic tsunami at different spatial resolutions. Computer simulations with resolution L/N, where L is the size of image scene (1000 km × 1000 km) and N is the number of pixels (N = 5000, 1000, 500, 400, 200, 100).

4.6 Data Fusion

More detailed investigations can be made using fusion of multiband imaging data, especially in the cases of superresolution microwave observations. We refer to this topic as *ocean remote sensing data fusion* (ODF). The goal of ODF is to improve the estimation of ocean microwave data and develop a tool for robust extraction of radiometric signatures. The material presented here is a part of our original work (Raizer 2013).

Multisensor and multispectral (MS) methods of data fusion are widely used in remote sensing (Alparone et al. 2015; Lillesand et al. 2015; Pohl and van Genderen 2016). Eventually, MS data carry more valuable information than data collected by single-frequency-band sensors—active (radar and lidar) or passive (radiometer, infrared, and video).

There are well-developed data fusion techniques and algorithms (Hall and McMullen 2004; Liggins et al. 2009; Tso and Mather 2009; Raol 2010); however, for ODF, we are focusing on the following three:

1. *Statistical method.* This pixel-level method provides the fusion of MS data using their statistical and correlation characteristics. Intensity–hue–saturation (IHS), local mean matching, principal component

analysis (PCA), regression analysis, statistical region merging, and other techniques are in most common use. This ODF method requires the collection of large bodies of statistical microwave data.

2. *Fast Fourier transfer (FFT) method.* This method provides low- and high-pass Fourier filtering (FFT) of raw data, fusion of filtered data, and the use of an inverse FFT (IFFT) to obtain the needed information enhancement. The FFT method is appropriate for ODF if one assumes that initial MS (imaging) data include periodic-like features. The best remote sensing example is the surface manifestation of oceanic internal waves (Chapter 5).

3. *Wavelet method.* As a pixel-feature-level fusion method, a multiresolution wavelet transform yields rich scale-dependent and structural information in both spatial and frequency domains. Data fusion is based on digital wavelet transform (DWT) that provides great enhancement of image features; DWT is widely used for the analysis of dynamic multiscale data. We assume that wavelet is the most efficient method for ODF.

An important practical topic is MS active–passive microwave imagery of the ocean surface with the highest spatial resolution (~1–10 m). Although such real-world remote sensing experiments are the greatest challenge thus far, nevertheless, the appropriate data fusion technique can be investigated using numerical simulations. In particular, the forward linear model for superresolution MS microwave imagery can be formulated in matrix-vector form, for example, according to (Nguyen et al. 2001):

$$\bar{\bar{T}}_A(k) = \bar{\bar{P}}(k)\bar{\bar{F}}(k)H_{atm}(k)\bar{\bar{T}}_{B0}(k) + \bar{V}(k), \quad k = 1,...,N \tag{4.7}$$

where $\bar{\bar{T}}_A(k)$ is the measured radio-brightness image scene for the k-th spectral band (sensor), $\bar{\bar{T}}_{B0}(k)$ is the actual high-resolution radio-brightness image scene, $\bar{\bar{P}}(k)$ is the sensor psf, $\bar{\bar{F}}(k)$ is the image sampling operator providing alignment of data with different spatial resolutions, $H_{atm}(k)$ is the atmospheric transfer function, $\bar{V}(k)$ is random noise, and N is the number of spectral bands (or sensors).

The formal solution of Equation 4.7 can be defined through the direct inverse technique:

$$\bar{\bar{T}}_{B0}(k) = [\bar{\bar{M}}(k)^T\bar{\bar{M}}(k)]^{-1}[\bar{\bar{M}}(k)]^T\bar{\bar{T}}_A(k), \tag{4.8}$$

where matrix $\bar{\bar{M}}(k) = \bar{\bar{D}}(k)\bar{\bar{F}}(k)H_{atm}(k)$. In the case of higher (super) spatial resolution, large dimensions of matrices $\bar{\bar{M}}(k)$ and $\bar{\bar{M}}(k)^T\bar{\bar{M}}(k)$ may significantly increase computation time and lead to unpractical results.

An initial ODF can be performed using the PCA method (Raol 2010), which for multiple data fusion can be formulated as a sum:

$$\bar{\bar{M}}(k_{ij}) = P_i \bar{\bar{M}}(k_i) + P_j \bar{\bar{M}}(k_j), \tag{4.9}$$

where $\bar{\bar{M}}(k_i)$ and $\bar{\bar{M}}(k_j)$ are input spectral images and P_i and P_j are normalized principal components (i.e., $P_i + P_j = 1$) computed from the covariance matrix, $k_{i,j}$ denotes spectral bands $(i,j = 1, 2, \ldots, N)$, and N is the number of spectral bands. The PCA is a standard pixel-level method of image fusion. The basic algorithm (4.9) is simple and available in MATLAB®; PCA provides relatively stable result in terms of image enhancement. However, implicit solution of Equations 4.7 and 4.8 using the maximum likelihood estimation algorithm and/or neural network (Benediktsson et al. 1990), is preferable for multiproposed ODF applications.

As a whole, the implementation microwave MS ODF is a challenging task because of low signal-to-noise and contrast-to-noise ratios of ocean microwave data. Indeed, the following reasons may affect the detection performance and feature extraction: (1) high correlation, anisotropy, nonuniformity, and nonstationary of MS ocean data that make it difficult to provide statistical matching and linkage; (2) typically low multispectral resolution that prevents obtaining valuable information in the frequency domain; and (3) a high level of environmental noise that usually poses a major challenge in the digital evaluation of relevant information.

However, the ODF process can be improved in some way if we organize parallel or distributed network based on combined (or hybrid) data fusion. We believe that the relevant signatures can be extracted and evaluated much better in the case of hybrid ODF than in the case of one selected data fusion method.

Figure 4.9 shows the suggested ODF algorithm. It consists of a number of formal operations (subblocks) providing multistep fusion between input MS data. Output product represents fused resulting data and extracted signatures. The main question occurring here is how to provide their geophysical characterization and specification? In our opinion, such an interpretation could be done using ancillary methods: numerical modeling, simulations, and *statistical matching* techniques.

Figure 4.10 demonstrates an example of a simple ODF realization. The algorithm is applied for a five-frequency set of multiband radiometric images generated digitally using a four-factor composite microwave emission model. The microwave contributions (brightness temperature contrasts) from surface roughness, foam coverage, surface temperature, and salinity are computed separately and incorporated into an imaging model stochastically.

Using certain multistep data fusion, it is possible to enhance the effects from surface roughness and foam coverage simultaneously to reduce the

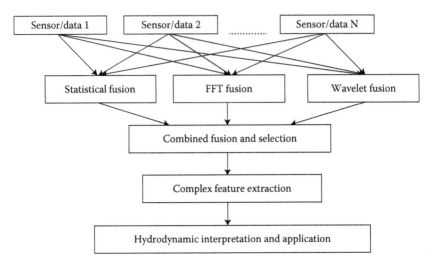

FIGURE 4.9
Multisensor/multiband data fusion network for advanced ocean remote sensing studies (ODF).

effects from temperature and salinity. This particular numerical example demonstrates how to extract or illuminate information from multiband microwave imagery using ODF.

Data fusion is an important part of practical applications, but the task is complicated for full implementation due to stochastic complexity and the multiple meaning of high-resolution MS ocean databases. This circumstance may lead to ambiguous interpretation of fused results. To provide the best result, a large volume of computations is required. As a whole, a combined (hybrid) data fusion (neural) network seems to have significant advantages for providing valuable assessment of MS ocean microwave data.

The ODF technique proposed can be applied in microwave remote sensing for the detection of many "critical" surface phenomena, including wave breaking and foam/whitecap fields, surface slicks, rip currents, and oil spills. In this case, a parallel and/or distributed data fusion network seems to have significant advantages for obtaining relevant information. Combined (hybrid) data fusion methods can also be useful for geophysical interpretation of microwave data, revealing of "hidden" information (signatures), and retrieval purposes as well.

4.7 Summary

In this chapter, we focused on a novel scientific topic concerning prediction, modeling, and simulations of complex ocean microwave remote sensing

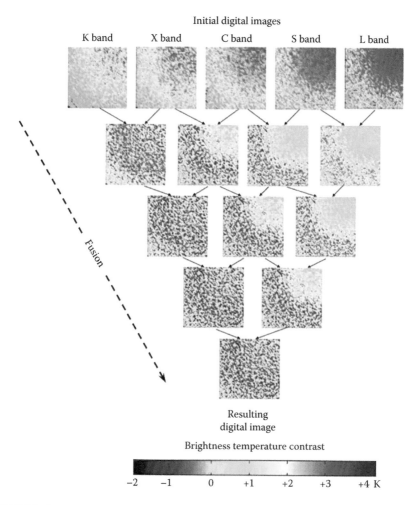

FIGURE 4.10

Digital example of multiband ODF fusion for microwave imagery.

data—signals, images, and signatures. For this goal, a combined digital framework, operated with multifactor microwave emission models and digital methods of data/image processing was developed and employed. During the numerical study, the following important practical issues are revealed:

- Joint multifactor effects and their statistical intermittency cause complexity and a variety of ocean microwave data that may lead to difficulties of their geophysical interpretation.

- In order to reduce possible uncertainties and provide an enhancement of the relevant information, flexible multispectral (multiband) fusion methods seem to be the most promising tool.

- The detection performance of ocean microwave signatures depend on an observation process (spatial resolution and averaging) significantly.
- Statistical multifactor models can provide the most realistic interpretation of complex radiometric signatures.

It is evident that the presented examples give us just an initial "quick" view on data content and signature prediction. Future developments that undoubtedly influence ocean remote sensing should include the use of efficient intelligent digital simulation techniques and algorithms (offering computer experiments), the creation of comprehensive physics-based imaging models, and providing purposeful field observations.

References

Alparone, L., Aiazzi, B., Baronti, S., and Garzelli A. 2015. *Remote Sensing Image Fusion (Signal and Image Processing of Earth Observations)*. CRC Press, Boca Raton, FL.

Benediktsson, J., Swain, P. H., and Ersoy, O. K. 1990. Neural network approaches versus statistical methods in classification of multisource remote sensing data. *IEEE Transactions on Geoscience and Remote Sensing*. 28(4):540–552.

Chen, C. H. 2007. *Signal Processing for Remote Sensing*. CRC Press, Boca Raton, FL.

Hall, D. L. and McMullen, S. A. H. 2004. *Mathematical Techniques in Multisensor Data Fusion*. Artech House, Norwood, MA.

Liggins, M. E., Hall, D. J., and Llinas, J. 2009. *Handbook of Multisensor Data Fusion: Theory and Practice*. CRC Press, Boca Raton, FL.

Lillesand, T., Kiefer, R. W., and Chipman, J. 2015. *Remote Sensing and Image Interpretation*, 7th edition. Wiley, Hoboken, NJ.

Myers, R. G. 2008. Potential for Tsunami Detection and Early-Warning Using Space-Based Passive Microwave Radiometry. Master's Thesis. Massachusetts Institute of Technology. Boston, MA. http://dspace.mit.edu/handle/1721.1/42913

Myers, R. G., Draim, J. E., Cefola, P. J., and Raizer, V. Y. 2008. A new tsunami detection concept using space-based microwave radiometery. In *Proceedings of International Geoscience and Remote Sensing Symposium*, July 6–11, 2008, Boston, MA, USA, Vol. 4, pp. IV-958–IV-961. Doi: 10.1109/IGARSS.2008.4779883.

Nguyen, N., Milanfar, P., and Golub, G. H. 2001. A computationally efficient image superresolution algorithm. *IEEE Transactions on Image Processing*. 10(4):573–583.

Pohl, C. and van Genderen, J. 2016. *Remote Sensing Image Fusion: A Practical Guide*. CRC Press, Boca Raton, FL.

Pratt, W. K. 2001. *Digital Image Processing*, 3rd edition. John Wiley & Sons, Inc., Hoboken, NJ.

Raizer, V. 1998. Microwave radiometric scenes and images of oceanic surface phenomena. In *Proceedings of International Geoscience and Remote Sensing Symposium*, July 6–10, 1998, Seattle, WA, USA, Vol. 5, pp. 2474–2476. Doi: 10.1109/IGARSS.1998.702250.

Raizer, V. 2002. Statistical modeling for ocean microwave radiometric imagery. In *Proceedings of International Geoscience and Remote Sensing Symposium*, June 25–26, 2002, Toronto, Canada, Vol. 4, pp. 2144–2146. Doi: 10.1109/IGARSS.2002.1026472.

Raizer, V. 2005. Texture models for high-resolution ocean microwave imagery. In *Proceedings of International Geoscience and Remote Sensing Symposium*, July 25–29, 2005, Seoul, Korea, Vol. 1, pp. 268–271. Doi: 10.1109/IGARSS.2005.1526159.

Raizer, V. 2009. Modeling L-band emissivity of a wind-driven sea surface. In *Proceedings of International Geoscience and Remote Sensing Symposium*, July 12–17, 2009, Cape Town, South Africa, Vol. 3, pp. III-745–III-748. Doi: 10.1109/IGARSS.2009.5417872.

Raizer, V. 2010. Simulations of roughness-salinity-temperature anomalies at S-L-bands. In *Proceedings of International Geoscience and Remote Sensing Symposium*, July 25–30, 2010, Honolulu, HI, USA, pp. 3174–3177. Doi: 10.1109/IGARSS.2010.5651356.

Raizer, V. 2011. Multifactor models and simulations for ocean microwave radiometry. In *Proceedings of International Geoscience and Remote Sensing Symposium*, July 24–29, 2011, Vancouver, Canada, pp. 2045–2048. Doi: 10.1109/IGARSS.2011.6049533.

Raizer, V. 2013. Multisensor data fusion for advanced ocean remote sensing studies. In *Proceedings of International Geoscience and Remote Sensing Symposium*, July 21–26, 2013, Melbourne, Victoria, Australia, pp. 1622–1625. Doi: 10.1109/IGARSS.2013.6723102.

Raol, J. R. 2010. *Multi-Sensor Data Fusion with MATLAB*. CRC Press, Taylor & Francis Group, Boca Raton, FL.

Szeliski, R. 2011. *Computer Vision: Algorithms and Applications (Texts in Computer Science)*. Springer, New York, London.

Tso, B. and Mather, B. 2009. *Classification Methods for Remotely Sensed Data*, 2nd edition. CRC Press, Boca Raton, FL.

5

High-Resolution Multiband
Techniques and Observations

In this chapter, the basic principles of high-resolution passive microwave observations of the ocean are considered. The performance of this technique is defined by many factors; the most important ones are instrumentation design technology, observation process, data analysis, and geophysical interpretation. The concept is illustrated and discussed with the use of selected experimental and numerical data, which represent a novel class of remote sensing information. This material gives the reader a greater ability to understand better passive microwave instrument capabilities, research methodology, and challenges of high-resolution observations of the ocean.

5.1 Introduction

Over the past several years, passive microwave radiometers have been successfully used for the remote sensing of the ocean and the atmosphere, facilitating the monitoring of surface temperature, salinity, near-surface wind vector, oil spills, boundary-layer characteristics, and air–sea fluxes. The sensitivity of radiometric measurements to ocean processes and parameters is quite different. It is a well-known fact that variations of microwave radiometric signals from the ocean are not only defined by surface conditions but also depend on the observation process. In order to obtain the desired information from remote sensing measurements, it is necessary to choose the appropriate technological configuration: microwave frequencies, instrument observing geometry, view angle (polarization), antenna footprint, spatial resolution, swath, and other parameters. All these motivations are defined by the goals and tasks of remote sensing studies.

Space-based passive microwave radiometric systems—SSM/I, SMMR, WindSat, SMOS, Aquarius (2011–2015), Aqua AMSR, Meteor-M MTVZA, and other microwave sensors—have been designed to facilitate the observation of Earth in low resolution (about 20–100 km), allowing for the monitoring of mesoscale and megascale geophysical parameters. Detailed descriptions of these and other satellite instruments and programs can be found in various books (Kramer 2002; Grankov and Milshin 2010; Ulaby and Long 2013;

Martin 2014; Qu et al. 2014) and papers (Kerr et al. 2001; Gaiser et al. 2004; Le Vine et al. 2007, 2010; Cherny et al. 2010; Klemas 2011).

Eventually, these missions are not productively efficient for the observation of localized dynamic ocean features. It is quite understandable for us that this task requires the application of more efficient remote sensing technology. In this chapter, we discuss the possibilities of high-resolution multi-band passive microwave imagery having considerable advantages in ocean remote sensing studies.

As mentioned above, the passive microwave technique is capable of providing detailed observations of the ocean surface in the case when a certain scientific methodology is applied. Two principal problems have arisen in this connection: (1) selection and evaluation of the relevant data, the so-called signatures of the interest and (2) their geophysical sense, validation, and correct physics-based interpretation.

An appropriate solution can be found using a combined theoretical–experimental approach. It means that the problem cannot be solved theoretically only, that is, without proper experiments or measurements; and conversely, an experiment cannot provide adequate understanding of the problem without the corresponding theory (although in our case, experiment is more preferable than theory). Thus, in order to obtain the relevant data, we have to employ a wide arsenal of available-at-the-present-time hardware and software techniques. This chapter gives a chance to realize this option.

5.2 Historical Background

Active remote sensing studies conducted in the past several decades reveal the possibilities for the detection of ocean surface features induced by different environmental processes and fields. Among the acting factors are wind vector variations, surface wave interactions and modulations, internal wave actions, surface currents, ship wakes, the generation of convective cells and turbulent vortexes in the ocean–atmosphere interface, and the other events.

These oceanic phenomena have been observed many times in airspace X, L, P, C, and Ku band radar images, beginning with the pioneering SEASET mission and works (Apel and Gonzalez 1983) and SAR Internal Wave Signature Experiment SARSEX (Gasparovic et al. 1988).

The first passive microwave imagery of the sea surface in high resolution was conducted by I.V. Cherny in the 1980s using an airborne scanning multichannel radiometer, operated at frequencies of 22.2, 31, 34, 37, 42, 48, 75, and 96 GHz. This instrument was equipped with a circular conical-scanning mechanism, providing observations of the surface with a view angle of 75° from nadir. Field (ship and airborne) experiments demonstrated microwave capabilities to observe different "critical" sea surface phenomena, environmental and induced events. They are the following:

1. Observations of internal waves and solitons
2. The "relic rain" surface effect
3. Frontal zone in Kuroshio region
4. Oceanic synoptic ring (Rossby soliton)
5. Surface effect from the origin of tropical cyclone Warren
6. Diagnostics of anomalous cyclone trajectory in North-Western Pacific

A detailed review of these experiments and obtained data has been published in our book (Cherny and Raizer 1998). As a result of these studies, a new concept of ocean microwave diagnostics, based on the so-called amplification mechanisms, has been suggested and developed. In this concept, the amplification process is associated with the development of the secondary modulation instabilities that leads to a continuous or burst-type excitation of the wave spectrum. This effect causes strong variations of microwave radiometric signals at selective microwave frequencies that can be explained by the resonance theory of microwave emission (Section 3.3.2).

In the early 1990s, A.J. Gasiewski developed and created a novel similar-looking passive microwave multichannel radiometer-imager (but with a different conical-scanning mechanism), which was named "polarimetric scanning radiometer" (PSR) (Piepmeier and Gasiewski 1996, 1997; Klein et al. 2002).

The PSR is a versatile airborne mechanically scanned imaging radiometer with channels at 10.7, 18.7, 21.3, 37.0, and 89.0 GHz, including both vertical and horizontal polarizations at each of these bands. A key feature of the PSR is the ability to provide both forward and backward mapping from the aircraft using full-conical (360°) azimuthal-angle scans (http://www.esrl.noaa.gov/psd/technology/psr/).

The PSR was originally developed for the purpose of obtaining polarimetric microwave emission imagery of Earth's oceans, land, ice, clouds, and precipitation. The PSR has provided a unique opportunity to study the spatial structure of the ocean microwave emission at scales of brightness and spatial variations below 1 K and 1 km. In particular, an airborne PSR was used successfully for the measurement of the near-surface wind vector (Gasiewski et al. 1997; Kunkee and Gasiewski 1997; Piepmeier et al. 1998; Piepmeier and Gasiewski 2001). The PSR system was also involved in several environmental remote sensing experiments and subsatellite track missions (Jackson et al. 2005; Cavalieri et al. 2006; Bindlish et al. 2008; Stankov et al. 2008). For our ocean studies, the PSR has been installed on two aircraft (NASA P-3B and DC-8) and operated during the period 1997–2004. The first information about this experimental work was reported by Raizer and Gasiewski (2000). Table 5.1 shows the specification of the PSR system at this time.

TABLE 5.1

Specifications of Polarimetric Scanning Radiometer (PSR, 1998)

Platform	DC-8 or P-3B				
Center frequency (GHz)	10.7	18.7	21.5	37	89
Wavelength (cm)	2.8	1.6	1.4	0.81	0.34
Polarization	Vertical and horizontal				
Incidence angle from nadir	65°, 62°, and 58°				
Integration time (ms)	18				
Measured sensitivity (K)	0.5	0.3	0.4	0.5	0.6
for $\tau = 18$ ms					
Absolute accuracy (K)	1–2				
Estimate signal stability (K)	0.6–0.9				
for 200-km track					
Antenna 3-dB beamwidth	8°	8°	8°	2.3°	2.3°
Observation altitude ("H," km)	~1.0–3.0				
Antenna footprint size at:					
58° incidence	0.36H = 1.1 km 0.11H = 0.32 km				
65° incidence	0.52H = 1.5 km 0.15H = 0.45 km				
Swath width at:					
58° incidence	3.2H = 9.6 km				
65° incidence	4.3H = 12.9 km				

Source: Piepmeier, J. P. and Gasiewski, A. J. 1996. Polarimetric scanning radiometer for airborne microwave imaging studies. In *Proceedings of International Geoscience and Remote Sensing Symposium,* May 27–31, 1996, Lincoln, Nebraska, Vol. 3, pp. 1120–1122. Doi: 10.1109/IGARSS.1996.516587; Raizer, V. Y. 2005b. High-resolution passive microwave-imaging concept for ocean studies. In *Proceedings of MTS/IEEE OCEANS 2005 Conference,* September 18–23, 2005, Washington, D.C., Vol. 1, pp. 62–69. Doi: 10.1109/OCEANS.2005.1639738.

5.3 Basic Concept

An idea to apply high-resolution passive microwave imagery for detailed observations of ocean surface features has been formulated by the author in 1996 and reported later (Raizer 2005a,b). It was clear from previous experiences that two-dimensional microwave radiometric realizations (the so-called microwave portraits or pictures), in addition to the same one-dimensional radiometric records (or profiles), have essential advantages. High-resolution airborne passive microwave imagery is an efficient tool for exploring the ocean environment and conducting testable scientific experiments; it is a great opportunity to obtain new results as well.

Indeed, manifestations of oceanic processes and fields are shown on unified two-dimensional images much better than through one-dimensional records. This is because the continuous scanning regime provides instantaneous registration of spatial motions and variability of the surface. Even a single-frequency microwave picture perfectly made in the right place at the right time can give us much more useful information than a set of multi-channel radiometric profiles or time series.

The best option, therefore, would be to conduct multiband panoramic imagery that provides a multispectral radio-brightness portrait of the ocean with the highest spatial resolution. Such an imaging concept, however, requires new technology efforts and resources, including a novel view on the remote sensing problem. Thus, the main objectives of high-resolution observations are formulated as follows:

- The use of remote sensing technique for manifestation of ocean features
- Collect and specify appropriate microwave imaging databases
- Explore properties and parameters of ocean microwave images
- Develop statistical structural and textural characterization of imaging data
- Apply robust digital processing for the extraction and evaluation of relevant information
- Create an appropriate physics-based imaging model
- Develop modeling and simulations of different ocean scenes and scenarios
- Formulate hydrodynamic hypotheses for the explanation of the observed signatures
- Provide correct geophysical interpretation and applications of microwave databases

Figure 5.1 illustrates this concept schematically. In fact, we have to deal with a multiple interdisciplinary framework, which is organized as a scientific research. In the next section, we consider some examples and explain how to utilize collected microwave radiometric data for advanced studies and applications.

5.3.1 Elements of Microwave Imagery

An imagery provides a unique opportunity to explore oceanic features through two-dimensional radio-brightness pictures. Variations of microwave emission registered by radiometer-imager reflect spatial dynamics and conditions of the surface; however, the quality of the imaging data depends on the scan geometry and instrument characteristics. At an airspace remote sensing measurement, the main factor is the relationship between sensitivity and resolution of the scanning radiometer. The choice and optimization of the instrument parameters is of fundamental importance in passive microwave observations of the ocean surface.

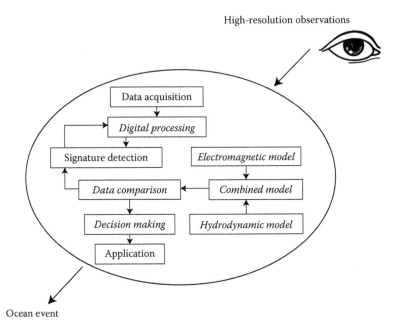

FIGURE 5.1
Basic microwave remote sensing concept for advanced ocean studies.

The geometry of aircraft observations is shown in Figure 5.2. In this case, the performance of microwave radiometric measurements or the so-called detection ability can be maximized at favorable environmental conditions. The detection ability is defined through the relationships between radiometric sensitivity δT, geometry of scanning measurements, and parameters of the deserved signatures $\{T_{con}, \eta\}$. In the case of a mechanically conical-scanned radiometer system, the detection ability D_T of the ocean target area can be estimated as

$$D_T = \frac{T_{con}\eta}{\delta T}, \tag{5.1}$$

$$\delta T = \sqrt{k^2(T_N + T_S)^2 \frac{1}{\Delta f \tau} + (T_N - T_E)^2 \left(\frac{\Delta G}{G}\right)^2} \quad \text{or} \quad \delta T = \frac{k(T_N + T_S)}{\sqrt{\Delta f \tau_{eff}}}, \tag{5.2}$$

$$\tau_{eff} = \frac{\beta / 2}{2\alpha} t = \frac{L_x L_y}{2DV}, \tag{5.3}$$

where δT is the fluctuation sensitivity of the radiometer; T_{con} is the brightness-temperature contrast of the target area; $\eta \leq 1$ is the beamfill factor; Δf

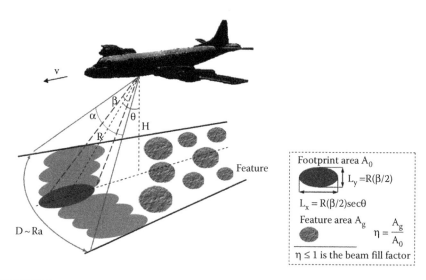

FIGURE 5.2
Geometry of high-resolution microwave imagery. (From Raizer, V. Y. 2005b. High-resolution passive microwave-imaging concept for ocean studies. In *Proceedings of MTS/IEEE OCEANS 2005 Conference*, September 18–23, 2005, Washington, D.C., Vol. 1, pp. 62–69. Doi: 10.1109/OCEANS.2005.1639738.)

is the frequency bandwidth (MHz); τ is the integration time of the radiometer; $\Delta G/G$ is the fractional power gain variation of the radiometer; T_N is the receiver noise temperature; T_S is the scene antenna temperature; T_E is the standard load noise temperature (for noise-injection radiometer $T_N - T_E \rightarrow 0$); the constant $k = \sqrt{2} \div 4$ depends on the type of the radiometer; τ_{eff} is the effective integration time (s), which depends on the parameters of scanning (also known as *dwell time per antenna footprint*); $\{L_x, L_y\}$ are the sizes of antenna footprint, that is, resolutions along the scan line and along the track line, respectively; $t = L_x V$ is the time of shift; θ is the incidence angle (constant); β is the antenna beamwidth (3 dB); α is the active scan angle; R is the distance; H is the altitude; D is the swath; and V is the speed of the aircraft (m/s).

Equations 5.1 through 5.3 have shown that reliable detection of regular radiometric signatures with absolute values $|T_{con}| \approx 1.5 \div 3.0$ K at $\eta \approx 0.5 \div 1.0$ can be performed if the following parameters are chosen: $\delta T \approx 0.15 \div 0.20$ K (for the frequency range $f = 10 \div 20$ GHz), $\Delta f = 300$ MHz, $T_N = 500$ K, $T_S = 200$ K, $\theta = 60°$, $\beta = 20°$, $\alpha = 120°$, $H = 3.0$ km, and $V = 120$ m/s. In this case, the values of the detection ability are in the range $D_T \approx 5 \div 10$, which is perfect for the registration of low-contrast ocean signatures with spatial sizes about 0.5–1 km.

The needed values of the detection ability D_T may not be reached if the quality (resolution and contrast) of the obtained microwave pictures is poor. Here, we set a contrast threshold $|T_{con}| \approx 3.0$ K, which is a critical physics-based level for the detection of background ocean parameters. It means that signatures with $|T_{con}| \geq 3.0$ K can be easily observed using an airborne real aperture microwave imager of standard configuration with a good-enough spatial resolution. Such signatures are usually associated with large-scale ocean surface dynamics.

However, it will be difficult and may be impossible to measure signatures with contrast $|T_{con}| \leq 1.5$ K using a conventional microwave technology. Such relatively low-contrast and slightly observable (hidden) signatures are usually associated with dynamic small-scale hydrodynamic disturbances, including deep-ocean surface manifestations as well. Their reliable detection requires the use of advanced technology and certain geometry of observations.

There are two options to improve the quality of passive microwave imagery in order to obtain the needed result. One possibility is a technical solution based on the use of either interferometric aperture synthesis or multiple-look real aperture radiometry with the highest spatial resolution. The principles of such microwave radiometric systems are known (Ruf et al. 1988; Le Vine 1999; Skou and Le Vine 2006). Another possibility is to use conventional scanning radiometric systems and develop and apply robust digital processing for the enhancement of microwave imaging data. In this case, the relevant information may be extracted and specified using image processing algorithms and computer vision techniques. We consider this option.

5.3.2 Elements of Digital Processing

Digital processing of microwave imaging data is an important part of high-resolution ocean observations. The goal of the processing is to provide selection, extraction, and evaluation of the relevant information (signatures) having geophysical sense. The processing includes a number of operations and manipulations related to statistical, textural, structural, and morphological analyses of imaging data. Such a processing also sets criteria and rules needed for signature identification and decision making.

Figure 5.3 shows a common scheme of the thematic data processing. This chart contains several blocks. The preprocessing (I) provides two-dimensional formatting of raw radiometric data, including geometric correction, instrument noise reduction, calibration, geographical positioning, and display of multiband images. Global processing (II) is applied for texture characterization, enhancement, and classification of geolocated images. Local processing (III) provides feature extraction, segmentation, and morphological analyses. Shape, size, and brightness are used for the selection and specification of relevant signatures. Statistical processing (IV) is applied additionally for estimating spatial, spectral, and correlation characteristics of images and signatures.

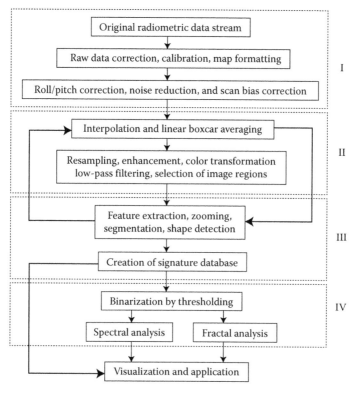

FIGURE 5.3
Flowchart of thematic image processing: (I) preprocessing; (II) global processing; (III) local processing; (IV) statistical processing.

5.3.3 Elements of Interpretation

The interpretation of ocean microwave imagery is a complicated process; it requires the use of an interdisciplinary approach and experiences. Interpretation is based not only on the understanding of ocean environment as a whole hydrodynamic system but also includes the knowledge of physical mechanisms and dependencies of ocean microwave emission. Interpretation also invokes modeling and simulations of microwave radiometric data.

Figure 5.4 illustrates a flowchart (so-called radio-hydro-physical model) that is used for the geophysical interpretation of high-resolution ocean microwave data. It includes three main parts: hydrodynamic (upper contoured block), electromagnetic (middle contoured block), and data utilization (lower contoured block).

The hydrodynamic part describes the mechanisms of generation and evolution of ocean features *potentially* detectable by microwave radiometer-imager. It does not mean that all available hydrodynamic theories and/or

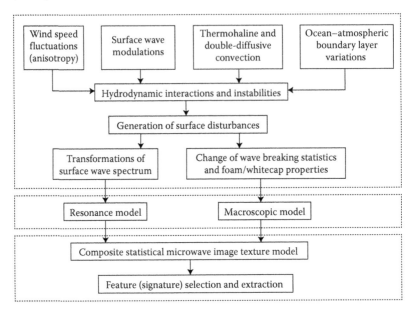

FIGURE 5.4
Hydrodynamic-electromagnetic model for analysis of complex data.

models should be involved in the process of interpretation. It is necessary to consider just mechanisms and effects responsible for the formation of measurable variations in ocean microwave emission.

The electromagnetic part involves modeling and simulations of spectral band and polarization characteristics of ocean microwave emission. To perform an adequate theoretical analysis, a multifactor approach is considered (Sections 4.2 and 4.4).

Finally, the utilization provides digital specifications of the signatures through the comparison between experimental and model prediction data. This process requires multiple operations to achieve the needed results.

5.4 Analysis of Microwave Data

The methods of digital data processing dedicated to the thematic analysis of high-resolution ocean microwave imaging data are considered. We develop and apply combined algorithms and techniques based on an experimental–theoretical approach that allows us to provide flexible data processing depending on the goal and scientific task. We believe that this option gives us the best possibilities and advantages for the correct interpretation and application of high-resolution ocean microwave data.

5.4.1 Imaging Data Collection

In this section, selected examples of experimental high-resolution ocean imaging data are presented and discussed. The collection is created on the basis of the airborne PSR measurements. All observations were made under calm and moderate surface winds, stable air–sea conditions, and mostly clear atmosphere. Experiments included radiometric mapping of approximately 150 × 50 km test areas at altitudes of 1.5, 3.0, and 5 km. The flight pattern usually consisted of five, four, or three closely spaced parallel flight legs over the open ocean. As a result, in each experiments, two-look ("front" and "back") images corresponding to mapping ahead and behind of the aircraft were collected, calibrated, and archived.

Visualized PSR data represent a 20-channel set of multiband, formatted, geolocated, and calibrated digital images (radiometric maps) generated simultaneously at five frequencies: 10.7, 18.7, 21.5, 37, and 89 GHz (λ = 2.8, 1.6, 1.4, 0.81, and 0.33 cm wavelengths, respectively); both horizontal and vertical polarizations, and backward- and forward-looking positions; the range of available constant grazing view angles is 58–62° from the nadir; the spatial resolution is ~100–500 m depending on the microwave frequency and flight altitude.

Figure 5.5 shows a typical multiband set of high-resolution geolocated PSR images of the ocean surface at moderate wind. These radiometric images represent extended and local radio-brightness fields with variations of the brightness temperature in the range approximately from −5 to +5 K. It is a potentially valuable source of novel information, which requires special analysis. Methods analysis and interpretation of these data will be considered below.

5.4.2 Signature Specification

The specification of image features (or radiometric signatures) is an important part of the processing. For digital specification, three main criteria are used: shape, size, and brightness of signatures. The algorithm is built on the basis of morphological analysis, sorting, and filtering. The filtering is implemented as an interactive classifier and includes the following main procedures: (1) linear and nonlinear filtering, (2) color segmentation, (3) texture analysis, and (4) enhancement of radiometric signatures by specified brightness gradients. The last operation is performed using a tunable red–green–blue (RGB) color filter. The RGB filtering provides pseudo-color visualization of the signature's geometry.

Figure 5.6 shows typical examples of image features (signatures), extracted from geolocated multiband PSR data after digital image processing (Raizer 2003). These features represent small-scale image objects (spots) of reduced brightness having "cold" centers inside and "hot" multicolor contours outside. The range of their brightness contrast inside is about 3–5 K and typical sizes are 2–5 km. These spot-type features are most representative in a statistical sense. Detailed investigations show that other more complicated

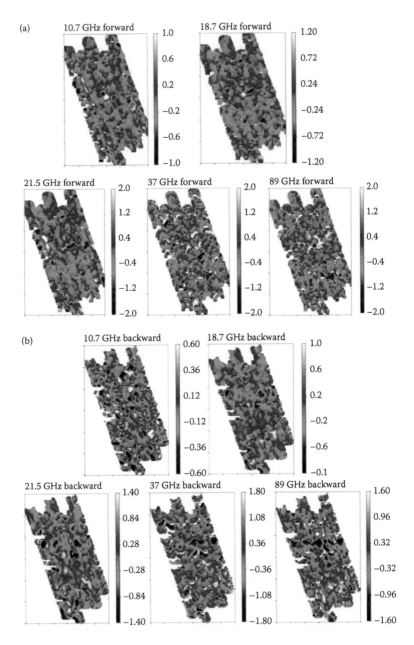

FIGURE 5.5

Multichannel set of high-resolution (~0.1–0.3 km) ocean PSR images. Five spectral channels are combined all together. $\theta = 62°$ incidence, horizontal polarization. (a) Forward look. (b) Backward look. Mapping area is ~20 × 30 km. Conditions: moderate wind, foam-free surface, and clear air. (From Raizer, V. Y. 2005b. High-resolution passive microwave-imaging concept for ocean studies. In *Proceedings of MTS/IEEE OCEANS 2005 Conference*, September 18–23, 2005, Washington, D.C., Vol. 1, pp. 62–69. Doi: 10.1109/OCEANS.2005.1639738.)

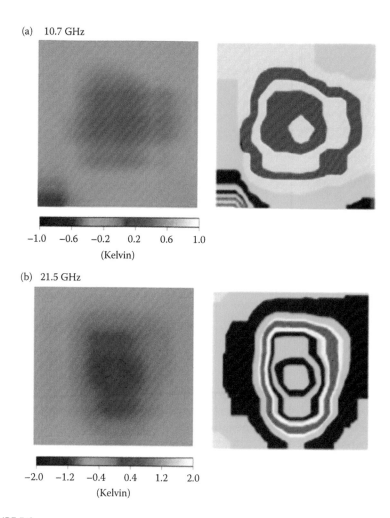

(a)　10.7 GHz

−1.0　　−0.6　　−0.2　　0.2　　0.6　　1.0
(Kelvin)

(b)　21.5 GHz

−2.0　　−1.2　　−0.4　　0.4　　1.2　　2.0
(Kelvin)

FIGURE 5.6
Example of spotted circle-type ocean microwave radiometric signatures and their color segmentation. (a) Cold spot, 10.7 GHz. (b) Hot spot, 21.5 GHz. Left part—true color, right part—segmentation. The size of all image fragments is ~0.8 × 0.8 km. (Raizer, V. Y. 2005b. High-resolution passive microwave-imaging concept for ocean studies. In *Proceedings of MTS/IEEE OCEANS 2005 Conference*, September 18–23, 2005, Washington, D.C., Vol. 1, pp. 62–69. Doi: 10.1109/OCEANS.2005.1639738.)

geometry features in the images (complex signatures) can be extracted as well. Some specific types of signatures are shown in Figure 5.7 and listed in Table 5.2. Potentially, the following suggestions can be made concerning their geophysical nature.

First, spotted radiometric signatures can be associated with a spatial nonuniformity of the near-surface wind. Surface slicks and/or rip currents are the most probable causes. Indeed, according to model calculations, the

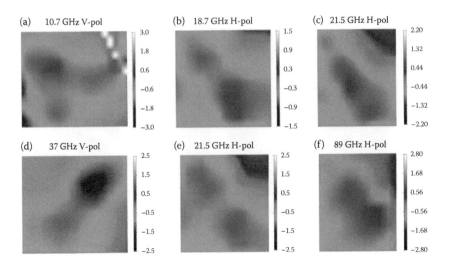

FIGURE 5.7
Selected examples of low-contrast ocean microwave radiometric signatures of variable geometry. (a) Cross-type. (b) and (c) V-type; (d)–(f) "Figure-eight"-type. The size of all image fragments is ~2 × 3 km or less. The color bar shows the brightness temperature contrast calculated relatively the mean level of the image intensity.

impact from the surface roughness at the selected PSR frequencies yields variations in the brightness temperature in the range ~–5 to +5 K at moderate (foam-free) conditions. Both negative and positive brightness temperature contrasts are observed at grazing view angles; polarimetric effects induced by the wind vector azimuthal anisotropy take place as well. The main hydrodynamic–electromagnetic factor here is transformations in surface wave spectrum, which produce measurable variations in ocean microwave emission (Chapter 3).

Second, radiometric signatures may reflect certain *thermohydrodynamic* conditions of the ocean upper layer (for example, the existence of subsurface

TABLE 5.2

Specifications of Distinct Radiometric Signatures

#	Type	Brightness Contrast (K)	Size (km)	Shape
1	Cold spot	–5.0...–0	<2...3	Circle, ellipse
2	Hot spot	0...+5.0	<2...3	Circle, ellipse
3	Figure "eight"	–3.0...–1.0	~2	Two closed spots
4	V-type	–2.0...+2.0	1.5...5	Fat cross
5	Tail-type	–2.0...+2.0	1.5...10	Fat line
6	U-type	–0.5...–1.5	<1	Horseshoe map

thermohaline fine structure) as a local energetically active zone. Such a "non-equilibrium" substance is favorable for the development of different types of hydrodynamical instabilities, generation of burst-type surface disturbances, (sub)surface turbulent intrusion, and even self-similar structures in the fields of surface wind and/or roughness.

Furthermore, some circle-type radiometric features observed in the high-resolution ocean images can also be microwave indicators of deep-ocean processes, which may cause the generation of mid-scale (~1 km) coherent structures. Among them, small eddies, vortexes, or turbulent cells at the air–sea interface are the most probable phenomena. We know that coherent ocean structures of different scales are visible in the satellite SAR images (Alpers and Brümmer 1994; Li et al. 2000; Ivanov and Ginzburg 2002).

During thematic analysis of the collected data, the following important conclusion can be made. High-resolution ocean microwave images represent multiple radio-brightness texture fields (mosaics) comprising spotted features of variable geometry and contrast. From a geophysics point of view, such microwave mosaic pictures are perceived as environmental "ocean microwave stochastic background" associated with the ocean–atmosphere interaction, including surface dynamics as well.

5.4.3 Texture Characterization

Texture-based algorithms are widely used in image modeling, segmentation, classification, pattern reconstruction, and computer vision. Texture increases the realism of the produced images, showing their fine structure and composition in great detail. Texture analysis was applied for investigations of ocean passive microwave imagery for the first time (Raizer et al. 1999).

Three principal methods can be used for the description of image texture: (1) structural, (2) stochastic, and (3) spectral. Structural techniques characterize textures as an arrangement of pixels, objects, or (sub)patterns according to certain placement laws. Stochastic techniques provide a global characterization of texture as a random field. The statistical properties of an image are determined by the probability density function (pdf). The spectral method describes the spatial regularity of texture features that can be investigated through Fourier analysis.

The most efficient option for the geophysical interpretation of high-resolution ocean microwave imagery is texture fitting or texture matching between experimental and model imaging data. Many texture models have been proposed after the pioneering publication using texture synthesis (Rosenfeld and Lipkin 1970). A review of the existing image processing methods (Gonzalez and Woods 2008; Li 2001; Richards and Jia 2005; Pratt 2007; Mirmehdi et al. 2008; Engler and Randle 2009; Mather and Koch 2011; Russ and Neal 2015) shows that it makes sense to consider a number of texture-based models. They are the following: (1) mosaic—cell and "bombing," (2) periodic, (3) Fourier series, (4) Brownian motion, (5) fractal, and (6) Markov chain. Models (1) and (2) are deterministic; models (3–6) are statistical.

Although all these models differ in the mathematical sense, the algorithm for image texture simulations can be formulated using a unified method. It is based on statistical characterization and quantification of a random radio-brightness field, generated as an ensemble of pixels with a certain pdf. An actual microwave image scene is represented by a two-dimensional discrete array:

$$T_B(x,y) = \sum_{n=1}^{N}\sum_{m=1}^{M} T_{Bn,m} w_{n,m}(x,y) J_{n,m}(x,y)\Delta x\Delta y, \qquad (5.4)$$

where $T_{Bn,m}$ is the pixel brightness temperature in point $\{n, m\}$ calculated from a microwave emission model, $w_{n,m}(x, y)$ is the corresponding weight coefficient related to the pdf (or histogram) of the chosen random field model, the kernel $J_{n,m}(x, y)$ represents the impulse response function of the linear image model, Δx, Δy are discrete sampling intervals, $\{n, m\}$ are current pixel indices, and $N \times M$ is the total number of generated pixels in an image. This technique has been applied for modeling and simulations of ocean microwave texture scenes with variable parameters (Raizer 2002, 2005a,b).

This numerical algorithm includes the following consecutive operations:

1. Generation of a discrete field of large numbers of pixels in a coordinate plane by certain deterministic or statistic laws.

2. Calculations of the pixel intensities, that is, the values of the brightness temperature (contrast) related to different oceanic factors. For this, microwave emission models or empirical approximations are used.

3. Labeling of the pixels covered specified image region, pattern, or geometrical objects. This procedure provides a preliminary coding of image features, which could be associated then with the signatures of interest.

4. Color quantization of the pixels by intensities. This operation arranges the value of the calculated brightness temperature for each pixel in standard RGB color format.

5. Digital interpolation, gridding, and sampling of the image scene or selected image fragments. These procedures provide the transformation of a quantitative discrete pixel field into a continuous color image of specified size.

During these operations, an actual two-dimensional microwave image $T_B(x, y)$ in terms of the brightness temperature (or contrast) is generated digitally. This image corresponds to a given microwave scene with specified parameters. Algorithm is applied at each radiometric channel separately (i.e., at each microwave frequency and polarization) using the corresponding theoretical approaches and assignments. As a result, a multidimensional array, that is, a multispectral digital microwave image, is generated.

For more realistic modeling, an observation process is invoked:

$$T_I(x,y) = \int P(x-x',y-y')T_B(x',y')dx'dy' + \Theta(x,y), \tag{5.5}$$

where $T_I(x, y)$ is the desired scene, $P(x', y')$ is the spread function describing the gain pattern of the radiometric antenna, and $\Theta(x, y)$ is the additive noise factor. Computations by Equation 5.5 require the knowledge of geometry of the experiment, spread function, and the signal-to-noise characteristics as well. Preliminary estimations show that in the case of low-contrast ocean microwave scenes (usually related to surface roughness), the observation process (5.5) does not change the resulting microwave image textures significantly. In the case of high-contrast ocean microwave scenes (for example, involving foam/whitecap coverage), ignoring the convolution (5.5) leads to an error or uncertainty at texture-matching image interpretation. Several digital examples of ocean microwave textures are shown in Figure 5.8. Similar textures can be found in the PSR images as well.

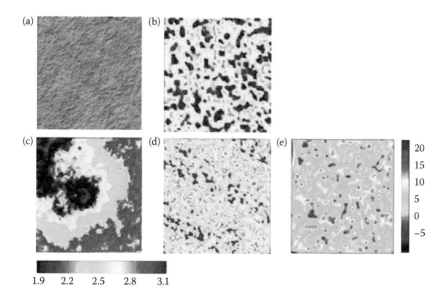

FIGURE 5.8
Stochastic ocean microwave radiometric textures simulated numerically using different random field models. (a) Initial image. Fourier series. (b) Mosaic. (c) Fractal. (d) Markov random field, MRF. (e) Combined multicontrast variable texture. Colorbars are shown in Kelvin. (From Raizer, V. Y. 2005b. High-resolution passive microwave-imaging concept for ocean studies. In *Proceedings of MTS/IEEE OCEANS 2005 Conference*, September 18–23, 2005, Washington, D.C., Vol. 1, pp. 62–69. Doi: 10.1109/OCEANS.2005.1639738.)

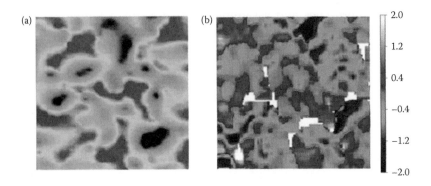

FIGURE 5.9

Example of statistical texture link between (a) model and (b) experimental image fragments (at 37 GHz) of the same size ~10 × 10 km. Colorbar is shown in Kelvin. (From Raizer, V. Y. 2005b. High-resolution passive microwave-imaging concept for ocean studies. In *Proceedings of MTS/ IEEE OCEANS 2005 Conference*, September 18–23, 2005, Washington, D.C., Vol. 1, pp. 62–69. Doi: 10.1109/OCEANS.2005.1639738.)

Figure 5.9 shows an example of a statistical link between model and experimental texture realizations of the same scales. The texture fitting method is used for this particular comparison. Numerical simulations are performed using a mosaic random field model operated with microwave characteristics of the ocean surface. It follows from this example that matching between model and experimental texture data can be realized, at least, qualitatively. This example shows stochastic matching only. Segmentation provides a more detailed comparison of local texture features.

The statistical properties of high-resolution ocean microwave images can also be investigated through a two-dimensional fast Fourier transform (2-D FFT). Fourier analysis is a tool providing information about the frequency content of a whole image as well as spatial distribution of image features. We apply this technique in order to estimate image texture regularity, searching for quasi-periodic variations of radio-brightness.

The original set of ocean microwave images and calculated power 2-D FFT spectra are shown in Figure 5.10. Most of the spectral features are seen to be concentrated in low- and mid-spatial frequencies: spectral domains are distinguished clearly enough using a color filtering of the 2-D FFT realizations. This processing yields the main spatial scales of low-contrast brightness variations in the images, revealing geophysical radiometric signatures. For these particular data, the relevant spotted signatures have a scale of ~2–4 km. Their angular distribution (i.e., peaks of radio-brightness) is observed in a wide sector of ~0°–60°, which may indicate the primary anisotropy of the surface.

The occurrence of stochastic mosaic textures and 2-D FFT spectral variations is a result of intermittent connectivity of microwave emission contributions induced by different environmental factors (Chapter 3). The joint electromagnetic impact produces the most realistic ocean microwave scenes,

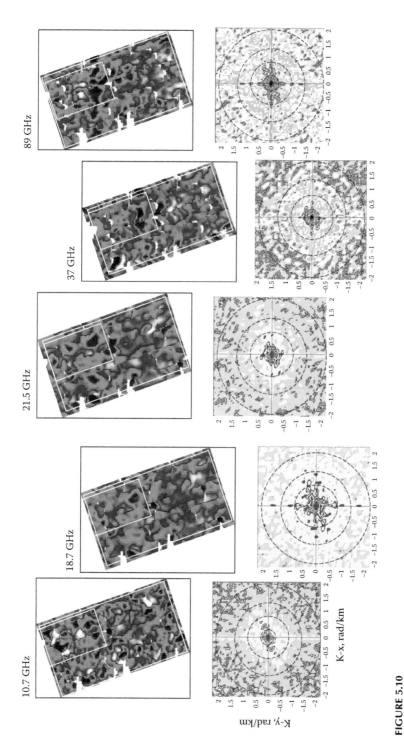

FIGURE 5.10
Enhanced Fourier spectra of ocean microwave radiometric images. Input transparencies are marked by white-line boxes.

which can be modeled and predicted, at least, statistically. For example, statistical combinations of microwave contributions from surface roughness and foam/whitecap result in the appearance of multicontrast stochastic microwave textures with randomly distributed distinct signatures (as shown in Figure 5.8e). However, involving any deterministic regular elements or geometrical objects into an imaging model will cause the image to change dramatically. A good example is a microwave portrait of internal wave events (Section 5.5.2). In this context, a method of texture characterization provides better insight into the content of ocean microwave imagery.

Operations with different deterministic and statistical texture realizations are known as a texture synthesis. Synthesized textures present *macrotextures* with complicated properties. Sometimes, macrotextures are useful for the recognition of complex features. Correlations between textures (multiband and spatial) may occur in the presence of quasi-periodic brightness variations in the images. Therefore, searching for correlated (and/or decorrelated) image texture features is necessary for the manifestation and specification of regular radiometric signatures.

5.4.4 Multiband Correlations

Correlation analysis is used to determine the structural and statistical characteristics of microwave imagery. This technique also gives us an important information about the sensitivity of different radiometric channels to geophysical parameters that is necessary for the optimization of multiband observations. Correlation-spectral formalism was developed and applied for the investigations of ocean PSR data (Raizer 2004).

A block diagram of combined spectral and correlation image processing is shown in Figure 5.11. The valuable spatial domains are defined using 2-D FFT spectra (examples are shown in Figure 5.10). For this, several operations are performed. First, the resampling procedure is applied for all multiband image transparencies, taking into account real scales of the image features. Then, 2-D FFT spectra of the selected transparencies are calculated and enhanced using digital image processing. Finally, both covariance and correlation matrices for different spatial domains are computed using the selected size of the image transparencies.

Statistical correlation properties of multiband PSR images are illustrated using two-channel correlation diagrams calculated directly from the calibrated raw data. Several examples are shown in Figure 5.12. The correlation

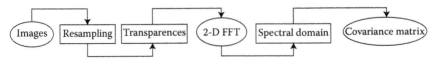

FIGURE 5.11
Block diagram of combined spectral and correlation image processing.

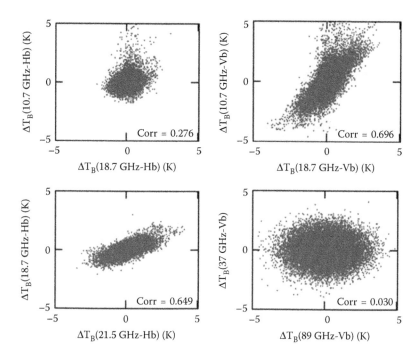

FIGURE 5.12
Typical two-channel correlation diagrams of ocean microwave radiometric data.

coefficient varies in a very large range approximately from 0.03 to 0.69. This method allows us to investigate texture correlations in a statistical sense and also to separate somehow the image regions with relatively weak and strong multiband correlations. However, from these common diagrams, it is difficult to estimate correlations between individual image features (signatures) and their groups.

A more detailed and adequate analysis involves computations of multiband covariance and correlation matrices. To demonstrate this technique, we employ a multichannel set of PSR imaging data. The following mathematical formulation is used:

$$
\left.
\begin{aligned}
&X = [x_1, x_2, \ldots, x_{n-1}, x_n]^T : && \text{Multichannel image with n bands} \\
&C_{xx} = E\{(X - m_x)(X - m_x)^T\} : && \text{Band-to-band covariance matrix} \\
&m_x = E\{X\} : && \text{Mean vector (the expectation)} \\
&R_x = \begin{bmatrix} r_{11} & r_{12} & \cdots & r_{1n} \\ r_{21} & r_{22} & \cdots & r_{2n} \\ \cdots & \cdots & \cdots & \cdots \\ r_{n1} & r_{n2} & \cdots & r_{nn} \end{bmatrix} : && \text{Correlation matrix}
\end{aligned}
\right\}
\tag{5.6}
$$

where x_i ($i = 1,2,...,n$) is the ith band data, $r_{i,j}$ ($i,j = 1,2,...,n$) is the correlation coefficient between the ith and jth band image, and T denotes the transpose.

Figure 5.13 shows examples of the covariance and correlation matrices computed by Equation 5.6 for ocean microwave imagery. These matrices are calculated for two selected types of the image fragments: large (a) and small (b). The fragments are marked in Figure 5.10 by two boxes (solid and dotted white lines, respectively). The covariance matrix reflects spatially statistical fluctuations of radio-brightness that is associated with frequency band dependencies of microwave emission. Environmental factors and stochastic properties of the surface produce approximately the same brightness temperature covariance at all microwave channels that lead to the apparent texture similarity of the images. Indeed, in the case of a large image fragment (a), there are relatively weak and mixed multichannel correlations with the coefficient 0.55–0.65, whereas in the case of a small image fragment (b), the correlation coefficient is 0.75–0.85 (for increased near-diagonal elements). The estimated correlation scale corresponds to periods of the main spectral components of the 2-D FFT spectra (Figure 5.10).

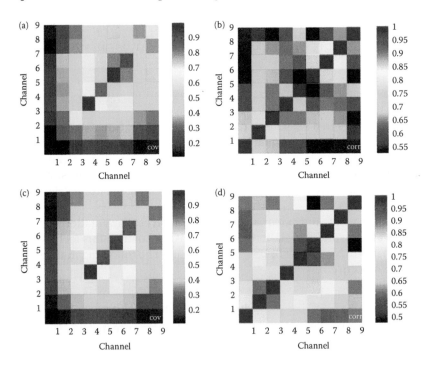

FIGURE 5.13
Multichannel covariance (a) and (c) "cov" and correlation (b) and (d) "corr" matrixes of size (9 × 9) for high-resolution ocean microwave images. Calculations were made for two different image fragments: large (a and b) and small (c and d) shown in Figure 6.10 by the corresponding boxes. Order of the PSR channels: 10.7h, 18.7h, 21.5h, 37h, 89h, 10.7v, 18.7v, 37v, 89v; channel 21.5v is missed; h, v—horizontal and vertical polarizations, respectively.

Multiband correlation between individual image objects or regions is an important measure providing additional statistical criteria for the manifestation of relevant signatures. However, it would be difficult to recognize and/or predict the signature appearance without hypothesizing about their origin or supporting information.

Correlation (decorrelation) characteristics of microwave signals indicate the occurrences of specific variations of geophysical parameters. For example, it could be transformations in the surface wave spectrum induced by fluctuations of the near-surface wind of other surface disturbances. Correlation-spectral analysis allows us to understand the situation better and provide more adequate interpretation of ocean microwave data.

5.4.5 Fractal-Based Description

Fractal geometry (Mandelbrot 1983) provides the simplest mathematical way to describe a scale invariance of natural objects. Unlike conventional geometry, fractal geometry deals with the shape's complexity through noninteger or fractal dimensions. Fractal objects can be found everywhere in nature, such as trees, flowers, ferns, clouds, rain, snow, ice, mountains, bacteria, and coastlines. Fractal geometry is also a tool for the investigation of nonlinear dynamical systems and complex phenomena with stochastic behavior: chaotic motions (attractors), random processes and fields, variable signals, and noises. In particular, a large number of literature recourses discuss geophysical fractals and fractal surfaces (Falconer 1990; Schertzer and Lovejoy 1991; Russ 1994).

In remote sensing, (multi)fractal techniques are used in order to characterize dynamic observations and experimental data, including multiresolution and hyperspectral images and features related to chaotic geophysical processes. In this context, fractal characteristics of ocean microwave imagery can be associated with certain *self-similar* hydrodynamic patterns. Fractal formalism allows us to develop a concept of "fractal signatures" that may provide robust detection and recognition of ocean features and events.

Fractal-based methods are involved for the description of sea surface dynamics and analysis of the induced optical, infrared, and scattered electromagnetic radiances (Glazman 1988; Glazman and Weichman 1989; Rayzer and Novikov 1990; Tessier et al. 1993; Kerman and Bernier 1994; Raizer et al. 1994; Shaw and Churnside 1997; Berizzi et al. 2004, 2006; Franceschetti and Riccio 2007; Sharkov 2007). By analogy with these works, we assume that under certain conditions, there occurs statistical self-similarity and scaling in ocean thermal microwave radiance as well.

One prominent environmental example is wave breaking and foam/whitecap activity at strong gales. Although the pronounced effect of foam/whitecap on ocean microwave emission is known (Chapter 3), however, spatially varying dynamical properties of wave breaking and foam/whitecap fields in the real world are not fully studied and described. Fractal-based techniques can provide the needed information from high-resolution optical and microwave images.

In both cases of ocean microwave and optical radiances, remarkable scale transformations are observed within the same intervals of surface wavelengths ~10–100 m. Spectral changes, scale invariance or scaling, occur due to cascade redistribution of energy within certain intervals of the wave spectrum. Therefore, high-resolution remote sensing measurements can sometimes yield self-similar realizations (signals, images, signatures) having specified values of the fractal dimension. In this sense, the registered microwave radiance can also be represented by (multi)fractal (Raizer 2001, 2012).

A time-dependent stochastic one-dimensional fractal signal can be modeled using the continuous wavelet transform (CWT) through the following reconstruction formula (Mallat 2009):

$$s(t) = \frac{1}{C_\psi} \int\limits_{-\infty}^{\infty} \int\limits_{-\infty}^{\infty} \frac{1}{a^2} [w(a,b)] \psi_{a,b}(t) da db, \quad \psi_{a,b}(t) = \frac{1}{\sqrt{a}} \psi_{a,b}\left(\frac{t-b}{a}\right), \quad (5.7)$$

where $w(a, b)$ are wavelet coefficients, $\psi_{a,b}(t)$ is the wavelet function,

$$C_\psi = \int\limits_{-\infty}^{\infty} \frac{|\Psi(\omega)|^2}{|\omega|} d\omega,$$

and $\Psi(\omega)$ is the Fourier transform of $\psi_{a,b}(t)$. Stationary scene-independent microwave radiometric signals usually correspond to a Gaussian process with Gaussian CWT coefficients. Because CWT is sensitive to non-Gaussian fluctuations dominating at some specific scales, the CWT is an effective technique in the wavelet synthesis of fractal stochastic signals.

In digitalized format, time series can be modeled using the following expression:

$$s(t) = \sum_\ell c_{j_0}(\ell) 2^{j_0/2} \varphi(2^{j_0} t - \ell) + \sum_{j=j_0}^{\infty} \sum_\ell d_j(\ell) 2^{j/2} \psi(2^j t - \ell), \quad (5.8)$$

where $\varphi(t)$ and $\psi(t)$ are the scaling and wavelet functions, respectively, $c_{j_0}(\ell)$ and $d_j(\ell)$ are scaling and wavelet coefficients, and j_0 is the integer value. Formula 5.8 allows simulations of deterministic and statistical signals depending on the choice of the scaling and wavelet coefficients. In the case of radiometric signals, variations of stochastic signal are calibrated by the brightness temperature using the r.m.s. fluctuation levels defined from model calculations (Chapter 3).

Fractal stochastic two-dimensional fields (surfaces and images) can be generated using the following well-known mathematical methods (Russ 1994):

- Fractional Gaussian noise (FGN)
- Fractional Brownian motion (FBm)

- Fractional Brownian motion by midpoint displacement (FBmMD)
- Wavelet synthesis (WLS)

In the case of FBm field, a stochastic fractal microwave imaging model can be represented as

$$\text{Cov}\{T_B(\vec{x})T_B(\vec{y})\} \propto \sigma_T^2 \{\| \vec{x} \|^{2H} + \| \vec{y} \|^{2H} - \| \vec{x} - \vec{y} \|^{2H}\}, \quad (5.9)$$

where $\text{Cov}\{T_B(\vec{x})T_B(\vec{y})\}$ is the covariance function of the brightness temperature, σ_T^2 is the variance, $\vec{x}$ and $\vec{y}$ refer to random coordinate vectors, and H is the Hurst exponent that is directly related to the fractal dimension D. The general relationship is $H = n + 1 - D$ for n-dimensional space. Covariance function (5.9) is written in the matrix form that allows us to simulate digitally microwave fractal images depending on the fractal dimension D.

In practice, scaling and self-similarity of microwave radiometric signals $S_T(t)$ and images $T_B(\vec{r}, t)$ can be investigated using the common scaling formulas:

$$S_T(\lambda t) = \lambda^D S_T(t), \quad (5.10)$$

$$T_B(\lambda \vec{r}, t) = \lambda^D T_B(\vec{r}, t), \quad (5.11)$$

where $T_B(\vec{r}, t)$ is the field of the brightness temperature (radio-brightness), λ is the scaling factor, D is the fractal dimension, $\vec{r} = \{x, y\}$ are the coordinates, and t is time.

On the other hand, the power spectral density of the radiance can be defined in standard form, $\Phi_T(f, t) = \mu(t)f^{-\beta}$, where β is the power exponent (or spectral index), f is the frequency, and $\mu(t)$ is the time-dependent amplitude. The linear scaling relationship between β and D over the range $1 \le D \le 2$ is $D = E + (3 - \beta)/2$, where E is the Euclidean dimension. This relationship is often used for the analysis of geophysical data sets; however, in the case of the ocean environment, a linear law may be not true due to nonstationary motions. It means that fractal characterization of ocean remote-sensed data can be performed at certain time-spatial frequency domains, where the spectral density itself does not change at all. For example, "microwave fractal portrait" may reflect somehow highly dynamic chaotic behavior of the surface at some specific conditions. Supposedly, it could be a turbulent wake or other localized hydrodynamic event.

Fractal-based digital framework is shown in Figure 5.14. It consists of three parts: hydrodynamic, electromagnetic, and data generation. The first part (I) provides a modeling of hydrodynamic phenomena or events. It may include self-similar solutions of the invariant forms of the Navier–Stokes equations or other analytical and numerical results. The kinetic equation can be invoked as well in order to describe the evolution of the wave spectrum. In the second part (II), the computation of microwave emission is performed

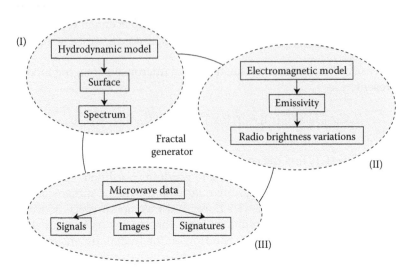

FIGURE 5.14
Framework diagram for modeling and simulations of fractal microwave data.

using electromagnetic models (Chapter 3). Spectral (multiband) and polarization (angular) dependencies of emissivity are parameterized by the fractal
dimension. The third part (III) provides simulations of microwave data and
their verification. Computer experiments allow us to explore complex microwave scenes and accomplish a multipurpose remote sensing task, including
fractal structurization of ocean microwave data.

Figure 5.15 shows examples of generated digitally fractal microwave radiometric signals and images. The performance is tested through the comparison between model and experimental radiometric data in terms of fractal
diagrams (Figure 5.16). Both data sets are processed using the traditional
box-counting method for computing the fractal dimension. It is assumed *a
priori* that an imaging set represents some kind of "microwave fractal." As
a result of routine operations, good correlations between model and experimental data can be achieved. In this case, self-similarity of ocean data is justified within limited scales (~1–3 km) and the values of fractal dimension
$D \approx 1.5$–1.9. As a whole, fractal properties of the images reflect mesoscale
ocean–atmosphere interactions involving hydrodynamic scaling as well.
Therefore, there occurs self-similarity in ocean microwave data obtained
under certain conditions.

Sometimes, the shape of individual radiometric signatures remains a morphological fractal or strange attractor. Figure 5.17 shows some examples of
fractal signatures (radio-brightness pictures) and estimates of the fractal
dimension. Figure 5.18 shows the comparison between the selected geometrical signatures and chaotic attractors: the hydrodynamic Lorenz attractor
and the *Smale* horseshoe map or horseshoe vortex (Smale 1967; Pesin and

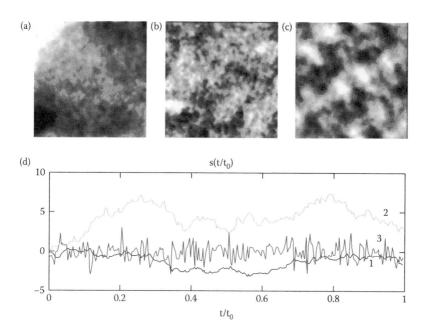

FIGURE 5.15
Fractal microwave images and signals. Modeling: (a) FBm. (b) FBm via midpoint displacement. (c) Fourier series. (d) Time-dependent fractal radiometric signal at Ku band. (1) Nonstationary (FBm) signal; (2) fractal anomaly (FBm via midpoint displacement); (3) stationary Gaussian noise (FGN).

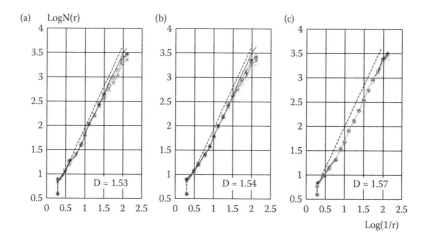

FIGURE 5.16
Fractal dimension computed from model and experimental microwave imaging data. Dotted lines—data from the modeling. Box-counting method is used: $D = \lim_{r \to 0}(\log N(r)/\log(1/r))$. Modeling for three variants (Figure 5.15): (a) FBm. (b) FBm via midpoint displacement. (c) Fourier series.

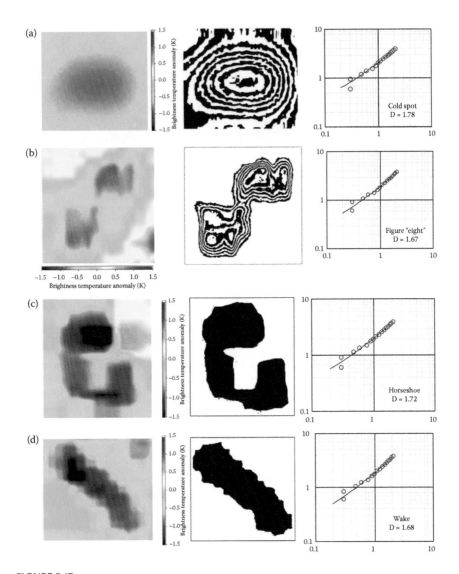

FIGURE 5.17
Radio-brightness signatures and fractal dimensions. Multilayer fractal: (a) cold spot; (b) "figure-eight." Morphological fractal: (c) horseshoe; (d) wake.

Climenhaga 2009). Note that both mathematical models are used for the description of hydrodynamic turbulence. The obtained result (Figures 5.17 and 5.18) demonstrates a good similarity between the values of fractal dimension computed for these types of geometrical signatures.

Fractal-like microwave signatures can also be manifested in the presence of mixing processes and turbulent intrusions (flows) in the upper ocean boundary (skin) layer. For instance, this situation may occur due to fluctuations of

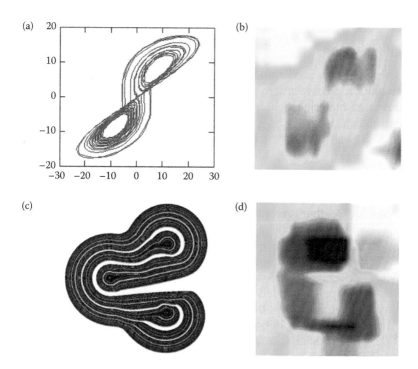

FIGURE 5.18
Fractal geometric radiometric signatures and chaotic attractors. (a) The Lorenz attractor and (b) related radio-brightness picture. (c) The *Smale* horseshoe map and (d) related radio-brightness picture. A horseshoe map is known as a mathematical realization of the turbulent horseshoe vortex system.

thermohaline fine structure in the field of internal waves and/or solitons (Chapter 2). The joint microwave impact of surface roughness, salinity, and temperature, the so-called roughness–salinity–temperature anomaly, can be detected using sensitive S–L band radiometers (Chapter 6).

Finally, the fractal dimension defined from the ocean microwave imagery and specified as a function of wind speed may become a distinctive "radiometric" equivalent of the Beaufort scale for sea surface state. We can see the analogy with optical data obtained from aerial photography.

Fractal-based methods have a significant advantage due to the possibility of involving scaling parameters in remote sensing models and data processing. As a whole, the apparent complexity of ocean high-resolution microwave data is associated with local nonuniformity, nonstationary, and intermittent events; their chaotic behavior; and fractual structure. These factors lead to the generation of fractal-like hydrodynamic features. Among them, natural surface slicks and rip currents, vortexes, eddies, wakes, *suloy*, breaking waves, foam and whitecap, oil spills, and other critical events are of a special interest for microwave remote sensing. In particular, remote sensing of *surface fractal anomalies* will be a valuable contribution to the development of the detection problem.

5.5 Observations

In this section, advanced remotely sensed data and results obtained using high-resolution passive microwave imagery are considered and discussed. The presented material includes observations of a wind-driven sea surface, oceanic internal waves, and ship wakes. This research demonstrates real capabilities of passive microwave imagery for advanced ocean studies.

5.5.1 Wind-Driven Sea Surface

Remote sensing measurement of the near-surface wind vector is an important application for weather forecasting, hurricane tracking, and marine services. The very first idea to retrieve wind speed from passive microwave radiometric data has been suggested in the late 1970s. The method was based on the measuring polarization anisotropy of ocean microwave emission (Dzura et al. 1992). Earlier references can be found in various papers (Pospelov 1996; Kuzmin and Pospelov 1999). During the last two decades, airspace methods of wind vector retrieval have been developed and applied (Wentz 1992; Yueh 1997; Krasnopolsky et al. 1995; Bettenhausen et al. 2006; Shibata 2006; Yueh et al. 2006; Colliander et al. 2007; Klotz and Uhlhorn 2014).

Meanwhile, reliable-enough estimations of environmental parameters for a wind-driven sea surface at scales less than 1–3 km always represent difficulties because of the low statistical representativity of remotely sensed data collected at localized areas. Moreover, short-term wind speed biases and wind stress cause an error in the retrieval of the near-surface wind vector at low winds. To improve estimates, the following technique is considered.

The microwave radiation–wind dependencies, that is, dependencies of the brightness temperature on wind speed, can be approximated using the so-called wind exponent. An idea is based on the assumption that spatial short-term variations (fluctuations) of wind speed under calm and moderated conditions produce distinct spotted radiometric signatures (as shown in Figure 5.6). These signatures—hot and cold spots—are distinguished in the images by geometric and brightness characteristics very well. Using statistical processing of the signatures related to variable wind conditions, the dependency of the brightness–temperature contrast $\Delta T_{Bs}(V)$ on wind speed can be defined. For this goal, the following approximate relation is used:

$$\Delta T_{Bs}(k_0, V) \approx 2T_0 k_0^2 \iint G(K, k_0; \varphi) F(K, V; \varphi) K dK d\varphi \approx 2T_0 k_0^2 \delta(k_0) V^\gamma \text{ or } \Delta T_{Bs} \propto V^\gamma,$$

$$(5.12)$$

where $\delta(k_0)$ is constant depending on the electromagnetic wave number k_0 and the dielectric parameters of sea water. Formula 5.12 is obtained from the resonance microwave model (Chapter 3) involving the power wave written in the form $F(K) = AK^{-4}Q(\varphi)$ at $A \propto V^\gamma$, where γ is the wind exponent and $Q(\varphi)$

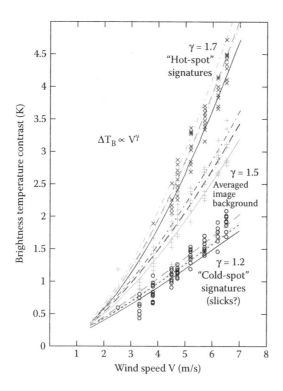

FIGURE 5.19

Retrieval of the wind exponent from PSR imagery using cold/hot-spot signature sorting and statistical processing. (From Raizer, V. Y. 2005b. High-resolution passive microwave-imaging concept for ocean studies. In *Proceedings of MTS/IEEE OCEANS 2005 Conference*, September 18–23, 2005, Washington, D.C., Vol. 1, pp. 62–69. Doi: 10.1109/OCEANS.2005.1639738.)

is the spreading function. Such a retrieval technique has been demonstrated previously (Trokhimovski and Irisov 2000).

Figure 5.19 presents an example of the retrieval of wind exponent. Digital processing of microwave imaging data and relationship (5.12) yield the values of the wind exponent $\gamma = 1.2$ and 1.7 for two different cases: cold spot (reduced roughness) and hot spot (increased roughness) at an observation angle 62° and horizontal polarization. These estimates are made using methods of signature sorting and distinctive criteria. A wind-exponent retrieval technique has some advantages over others because it provides quick estimates at weaker wind speeds (<5–7 m/s) in restricted ocean areas.

5.5.2 Internal Wave Manifestations

Remote sensing of ocean internal waves is still a challenging task for passive microwave radiometry and imagery. This topic is discussed here as a historical essay covering the period 1981–2001. The data collected during this time are limited but impressive. Experiences show that the surface manifestations of environmental internal waves—microwave radiometric signatures (MSIW)—are registered in the form of quasi-periodic variations of the brightness temperature, correlated with internal wave periods. Perhaps, after 2001, follow-up passive microwave observations of MSIW were not conducted. Meanwhile, such multiband radiometric measurements have a great value because they provide testing materials needed for the experimental verification of detection techniques.

The first microwave radiometric observations of MSIW were made in 1981–1985 from research vessels equipped with three-axis gyrostabilized platforms. Early data and references can be found in our book (Cherny and Raizer 1998) and the paper (Baum and Irisov 2000). Then, in 1992, the Joint U.S./Russia Internal Wave Remote Sensing Experiment (JUSREX'92) was organized and conducted in the New York Bight, Atlantic Ocean (Chapman and Rowe 1992; Gasparovic et al. 1993; Bulatov et al. 1994; Gasparovic and Etkin 1994). JUSREX'92 was the first (and last) postcold-war international mission, in which different airspace and shipborne active/passive microwave and optical sensors and also oceanographic measurements were employed all together in order to provide comprehensive studies of ocean internal waves. Later, some radiometric data were obtained from the blimp during the COPE'95 (Kropfli et al. 1999). In 2001, a novel experiment was conducted in the New York Bight again using airborne PSR observations. During this experiment, multiband passive microwave imagery (mapping) of internal wave manifestations was conducted for the first time. Some results were published (Raizer 2007).

5.5.2.1 Joint U.S./Russia Internal Wave Remote Sensing Experiment (JUSREX'92)

The JUSREX is still the most advanced and informative multisensor field ocean project of the twentieth century. The goal of this project was formulated as the demonstration of remote sensing capabilities to detect deep-ocean processes particularly associated with environmental internal wave dynamics. The JUSREX was conducted in the test area of the eastern end of Long Island, Atlantic Ocean, where hydro-physical characteristics of internal waves are known very well. The experiment was organized principally by the Johns Hopkins University Applied Physics Laboratory (JHU/APL) and Space Research Institute, Moscow (IKI).

The JUSREX was the first multisensor mission for the exploration of internal waves using passive/active microwave and optical methods of remote sensing.

The JUSREX instrumentation included U.S. and Russian SAR satellites (ERS-1 and Almaz-1), airborne (Tu-134 SKh, P-3, DC-8) Ku, X, C, and L band radars, high-resolution airborne optical camera MKF-6, and multifrequency microwave radiometers and scatterometers. Shipborne oceanographic and meteorological instruments were also employed to measure *in situ* surface parameters simultaneously with remote sensing observations. In fact,

TABLE 5.3

Specification of Airborne Tu-134 SKh Ku Band SLAR

Parameter	Value
Operating frequency	13.3 GHz (λ = 2.25 cm)
Transmitted power (peak)	60 kW
Transmitted pulse width	110 ns
Receiver bandwidth	16 MHz
Receiver sensitivity	−99 dB
Antenna beamwidth (azimuth)	0.0035 rad
Antenna dimensions	0.44×6 m
Swath width	12.5 km (H = 2 km)
Average geometric resolution	25 × 25 m
Pulse repetition frequency	2 kHz
Polarization	VV, HH
Aircraft velocity	100...160 m/s
Number of integrated samples/pixel	180, nominal; function of velocity
Sampling rate	6 MHz × 8 bits
Number of pixels/rows	512/512

Source: Cherny I. V. and Raizer V. Yu. *Passive Microwave Remote Sensing of Oceans.* 195 p. 1998. Copyright Wiley-VCH Verlag GmbH & Co. KGaA. Reproduced with permission.

this experiment pioneers the investigations of ambient hydrodynamic–electromagnetic processes, associated with internal surface wave interactions, dynamics of induced surface currents, and the influence of stable/unstable atmosphere conditions.

An important material was collected from the Soviet aircraft laboratory Tupolev Tu-134 SKh (registration number CCCP-65917). This aircraft carried several sensors: Ku band real aperture side-looking radar (SLAR), multifrequency set of microwave radiometers, and six-band aerial photo camera. Tables 5.3 through 5.5 show airborne instrument specification. Figures 5.20

TABLE 5.4

Specifications of Airborne Microwave Radiometers (1986–1992)

Device	Frequency (GHz)	Wavelength (cm)	Δf (MHz)	ΔT (K) $\tau = 1$ s	Antenna Beamwidth
R-18	1.6	18.6	125	0.10	30°
R-8	3.9	8.0	210	0.07	15°
RP-1.5 (3-channel polarimeter)	20.0	1.5	2000	0.15	9°
RP-0.8 (3-channel polarimeter)	37.0	0.8	1600	0.15	9°

Source: Cherny I. V. and Raizer V. Yu. *Passive Microwave Remote Sensing of Oceans.* 195 p. 1998. Copyright Wiley-VCH Verlag GmbH & Co. KGaA. Reproduced with permission.

TABLE 5.5

Specifications of the Six-Channel Multispectral Aerial Photo Camera MKF-6M

Channel #	1	2	3	4	5	6
		Four Visible			Two Infrared	
Spectral band	480 nm	540 nm	600 nm	660 nm	720 nm	840 nm
Focus of objective, f	125 mm					
Maximum optical resolution	150 lines/mm (~2–3 m for ocean conditions)					
Maximum relative aperture	1/4					
Field of view size	0.4–0.64 H (H=altitude, km)					
Size of photo frame	56 × 81 mm					
Basic altitude	3 and 5 km					
Scale (L = H/f)	From 1:20,000 to 1:40,000					
Overlap	20%, 60%, 80%					
Product	Photo film roll containing up to 2500 frames					

and 5.21 illustrate the aircraft laboratory and SLAR imaging geometry. Alternate horizontal (H) and vertical (V) polarization radio impulses were transmitted from two suspension antennas, located under the fuselage. Backscattering signals of the same polarization were received simultaneously and produce four separated radar images (HH, VV, HV, and VH polarization). The SLAR swath was about 13 km on each side of the aircraft at an altitude of 2 km (Figure 5.21a). Flight legs were usually 50…70 km. In total, seven flights from the NASA Wallops Flight Facility, Virginia were made to provide the registration of ocean internal wave signatures.

FIGURE 5.20
The aircraft laboratory Tupolev-134 SKh equipped by Ku band SLAR, multifrequency set of microwave radiometers, and MKF-6 six-band optical aerial photo camera. *Source:* http://russianplanes.net/id34257

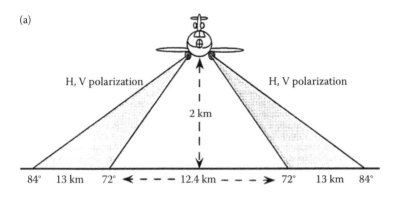

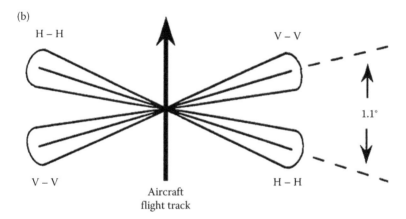

FIGURE 5.21
Diagrams of SLAR imagery (a) and antenna squint (b). (Adapted from Gasparovic, R. F. et al. 1993. *Joint U.S./Russia Internal Wave Remote Sensing Experiment: Interim Results*. JHU/APL Report S1R-93U-011. The Johns Hopkins University Applied Physics Laboratory, MD; Cherny I. V. and Raizer V. Yu. *Passive Microwave Remote Sensing of Oceans*. 195 p. 1998. Copyright Wiley-VCH Verlag GmbH & Co. KGaA. Reproduced with permission.)

The most important aspect of the SLAR imagery was the extreme sensitivity of the VV polarization signal to the atmospheric boundary layer stability (Gasparovic et al. 1993; Gasparovic and Etkin 1994). Figure 5.22a shows the reconstructed radar image mosaic designed from the SLAR data (Etkin et al. 1994).

Under stable atmospheric conditions, when the air temperature is higher than the surface water temperature, SLAR images on the vertical and horizontal polarization are qualitatively similar. The radar signatures are also similar, although the contrast due to internal waves on the VV polarization image is generally less than the contrast on the HH polarization image. Another picture is observed in the case of unstable atmospheric conditions. The HH polarization image at the top shows distinct internal wave

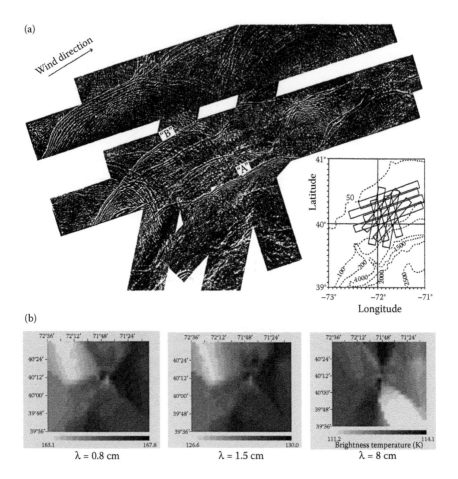

FIGURE 5.22
Remote sensing of ocean internal wave manifestations. (a) SLAR image mosaic and aircraft track. (b) Reconstructed passive microwave maps using three radiometric channels: $\lambda = 0.8$, 1.5, and 8 cm. Digital interpolation of one-dimensional radiometric records correspond to five flight legs (rosette) track pattern. JUSREX, July 21, 1992. (Adapted from Gasparovic, R. F. et al. 1993. *Joint U.S./Russia Internal Wave Remote Sensing Experiment: Interim Results.* JHU/APL Report S1R-93U-011. The Johns Hopkins University Applied Physics Laboratory, MD; Cherny I. V. and Raizer V. Yu. *Passive Microwave Remote Sensing of Oceans.* 195 p. 1998. Copyright Wiley-VCH Verlag GmbH & Co. KGaA. Reproduced with permission.)

signatures. In the VV polarization image, a cellular-type structure masks the internal wave signatures in the lower part of the image. The spatial scale of the cellular is a few kilometers. At the HH polarization image, the internal wave signatures are again presented.

Similar modulation-like signals and pictures were observed by other JUSREX radar sensors as well. Long-standing radar and *in situ* observations (Shuchman et al. 1988; Porter and Thompson 1999) as well as theoretical

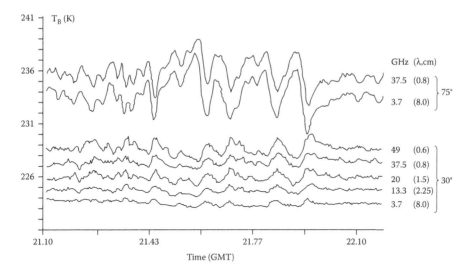

FIGURE 5.23
Passive microwave manifestations of ocean internal waves in the New York Bight, Atlantic Ocean. Multifrequency (right column, GHz) measurements at vertical polarization and 30° and 75° incidence angles from the *Academik Ioffe* research vessel during JUSREX'92. (Adapted from Bulatov, M. G. et al. 1994. In *Proceedings of International Geoscience and Remote Sensing Symposium*, Vol. 2, pp. 756–758.)

analysis (Liu 1988; Thompson et al. 1988) demonstrate stability and reproductivity of internal wave characteristics in the New York Bight.

Typical one-dimensional radiometric MSIW represent time-dependent oscillations of the ocean surface brightness temperature. Figure 5.23 shows radiometric records obtained *in situ* from the Russian *Academik Ioffe* research vessel during JUSREX (Table 5.6 shows shipborne instrument specification). Strong correlation and anticorrelation of radiometric signals were registered due to isopycnal and isothermal displacements associated with ocean internal gravity solitary waves.

Another situation occurred with the aircraft's microwave radiometers. A set of airborne radiometric data was used primarily for the reconstruction of sea surface temperature (SST) and wind speed vector. Using the principle of "polarizational anisotropy" (Chapter 3) and semiempirical regression algorithms, one-dimensional distributions of SST and wind speed vector along each flight leg were obtained.

Unlike shipborne data, no spatial modulation of radiometric signals due to the internal wave's effect was found. Only in the case of large-scale cellular features induced by atmospheric instability, some modulations of radiometric signals were observed. It was clearly an inconsistency between scales of internal waves and spatial resolution of the aircraft's radiometers (which was a few kilometers). At this averaging, only a large-scale change of integral characteristics of the ocean surface features can be measured by microwave

TABLE 5.6

R/V *Academik Ioffe* Radiometer Specifications (1992)

#	Device	Frequency F (GHz)	λ (cm)	Δf (MHz)	Sensitivity δT_{min}(K) $\tau = 1$ s	View Angle (degree)	Polari- zation	Antenna Beamwidth (degree)
1	Rp-0.6h	49.0	0.6	3000	0.06	25...80	H	8
2	Rp-0.8v	37.0	0.8	2000	0.15	25...80	V	8
3	Rp-1.5v	20.0	1.5	2000	0.20	25...80	V	9
4	Rp-1.5h	20.0	1.5	2000	0.20	25...80	H	9
5	Rp-8v	3.7	8.0	200	0.07	25...80	V	15
6	Rg-0.8v	37.5	0.8	1500	0.15	75	V	8
7	Rg-8v	3.7	8.0	500	0.13	75	V	9
8	Rs-0.5	60.0	0.5	3000	0.07	Scanning 270	Variable	5
9	R-IR	Infrared	8...12 μm	–	0.10	10	–	5

Rp: radiometer's view the sea surface at incidence angles of 25°–80°.
Rg: radiometer's view at 75°.
Rs: radiometer for scanning the surface and sky.

radiometers. In this case, no local surface effects, induced by single soliton or internal wave packets, can be manifested, as opposed to the shipborn's microwave observations.

An alternative method of the radiometric data processing was developed using two-dimensional statistical interpolation algorithm. The processing was applied for the spatial reconstruction of full microwave radiometric images of the test area where packets of internal waves were observed on the radar images. The main principle of the processing lies in choosing a two-dimensional low-frequency filter (or smoothing window) for the determination of large-scale microwave features associated with internal waves. Such a procedure was realized for multifrequency radiometric data set collected from our aircraft. As a result, spotted radiometric signatures were found in the interpolated images (Figure 5.22b).

The environmental conditions during JUSREX were described laconically by R. Gasparovic from The Johns Hopkins University Applied Physics Laboratory in the JUSREX'92 report: "During summer months, the water column in this area has three distinct layers: a thin mixed layer from the surface to a depth of about 10 m; a strongly stratified region from 10 to 25 m depth; and a weakly stratified lower layer extending to the bottom. Packets of internal waves are generated in the strongly stratified region by semidiurnal tidal flow over the shelf break. These wave packets propagate to the northwest and eventually dissipate when the water depth becomes less than about 25 m." The characteristics and periodicity of internal wave events are shown in Tables 5.7 and 5.8 (Jackson and Apel 2004). As follows from these and

TABLE 5.7

Characteristic Scales for the New York Bight Solitons

Packet Length (km)	Along Crest Length C_I (km)	Maximum Wavelength λ_{MAX} (km)	Internal Packet Distance D (km)
1...10	10...30	1.0...1.5	15...40
Amplitude $2\eta_0$ (m)	Long wave speed c_0 (m/s)	Wave period (min)	Surface width ℓ_1 (m)
−6 to −20	0.5–1.0	8...25	100

Source: Data from Jackson, C. R. and Apel, J. R. 2004. Prepared under contract with Office of Naval Research. Code 322PO. Internet http://www.internalwaveatlas.com/Atlas_index.html.

TABLE 5.8

Months When Internal Waves Have Been Observed in the New York Bight

Jan	Feb	Mar	Apr	May	Jun	Jul	Aug	Sep	Oct	Nov	Dec
			X	X	X	X	X	X			

Source: Data from Jackson, C. R. and Apel, J. R. 2004. Prepared under contract with Office of Naval Research. Code 322PO. Internet http://www.internalwaveatlas.com/Atlas_index.html.

other sources, solitary internal waves occur mostly in summer when heating enhances the stratification of the upper ocean. The solitons are generated by tidal flow near the edge of the continental shelf and occur in groups separated by some 20–35 km; amplitudes of 5–25 m and wavelengths from 200 to over 1000 m have been measured.

5.5.2.2 PSR Observations, 2001

JUSREX demonstrates very clearly that in order to register reliably passive MSIW, it is necessary to apply the high-resolution radiometric imaging technique. After 10 years, a novel remote sensing experiment has been organized and conducted in the same area in July–August 2001. For the first time, an airborne PSR imager was used for this ocean study.

Figure 5.24a shows a multiband set of experimental geolocated passive microwave images of the test area. These data were obtained at low wind and clear atmosphere in the New York Bight. In the PSR images, MSIW represent long low-contrast stripes, extended spotted regions, and short lines. Weak quasi-periodic variations of radio-brightness are visible as well. Moreover, there are some correlations between MSIW registered at different PSR channels. Additionally, the RADARSAT SAR satellite of the same ocean area (and approximately at the same time) was invoked in order to demonstrate the existence of internal wave packets and solitons (Figure 5.24b). Certain similarity between microwave radiometric and radar signatures of

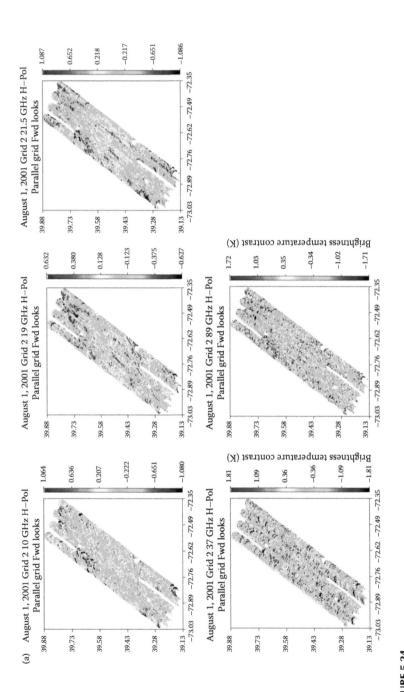

FIGURE 5.24
High-resolution passive microwave radiometric imagery of ocean internal waves in the New York Bight, Atlantic Ocean. August 1, 2001. (a) Five-band set of geolocated (latitude vs. longitude) microwave PSR images. Color maps ~80 × 20 km compound five parallel aircraft passes at 3000 ft flight altitude, 58° incidence angle, horizontal polarization, forward look. Radiometric signatures (red contrast stripes) are revealed well at 10, 19, and 21.5 GHz.
(Continued)

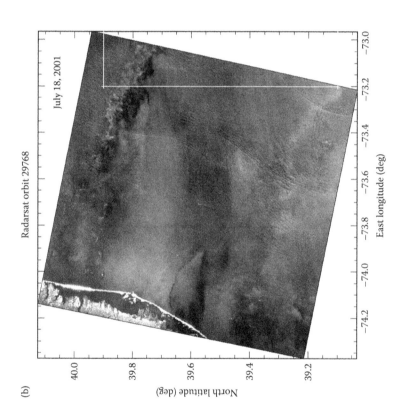

FIGURE 5.24 (*Continued*)

(b) Radar signatures of the same internal waves in the New York Bight, July 18, 2001. Marked region corresponds to PSR imagery. (RADARSAT image courtesy of D. Thompson, JHU/APL.)

internal waves is observed; both have the same band-stripe-type configuration, quasi-periodic spatial structure, and relatively low contrast.

Comparisons of shipborne and airborne radiometric data confirm the oscillation character of radio-brightness; however, the measured radiometric contrasts are different: 3–5 K (ship) versus 1–2 K (aircraft). This disagreement is explained not only by geophysical causes but also by significant differences in instrument resolution. For example, nonscanning shipborne JUSREX'92 radiometric measurements were made with a spatial resolution of 10–20 m and better, whereas aircraft scanning radiometric measurements had resolutions of about 100–200 m. Observation conditions and spatially temporal averaging of radiometric signals for these platforms differ as well.

An important part of this research is a validation based on real-world experiments. Although a detailed consideration of the internal wave theory is beyond the scope of this book, however, the most important physical models and mechanisms suitable for remote sensing studies make sense to list briefly. They are the following:

1. *Modulations of wind waves by gravity internal waves.* Strong internal waves reach the surface and cause the occurrence of surface roughness patterns—slicks and rip currents. These large-scale phenomena can be monitored by radar, radiometer, or optical camera.

2. *"Blocking" effects and wave cascades.* This mechanism is based on the Hughes theory. Kinematics of surface waves is determined by the group synchronism criterion: the sum of group velocity of surface waves and the velocity of the induced surface currents is equal to the phase velocity of the internal waves. In such a situation, surface waves propagating in the direction of strong opposing currents can be blocked by the current (Basovich and Tsimring 1984). Supposedly, blocking effects can enhance microwave backscattering and change emissivity.

3. *Nonlinear wave interactions, instabilities, and bifurcations.* This mechanism is based on a nonlinear wave theory formulated by Hasselman, Longuet-Higgins, and Zakharov; for more details, see Yuen and Lake (1982). Important remote sensing applications and theoretical studies were developed by Volyak; some results were published in Bunkin and Volyak (1987). This work was done, by the way, in order to explain probable occurrences of specific radar signatures in the presence of a moving submarine. Among them, a "cross," "fore-cursor," "arc," and other "nonlinearities" were the most abundant signature configurations. These data have been obtained in the late 1970s and 1980s using airborne Ku band side-looking two-polarization radar "Toros" (Antonov An-24 aircraft) operated at 2.25 cm wavelength.

4. *Effects of surface-active films.* This problem has been studied by many authors, for example, Gade et al. (1998) and Ermakov et al.

(1998). Internal waves accumulate the surfactant in the convergence zones and damp short gravity and capillary surface waves. This leads to transformations and modulations of wave number spectrum and corresponding variations in electromagnetic scattering. Theoretically, weak variations of emissivity can be registered as well, although it is difficult to control this effect in ocean experiments.

5. *Thermohaline circulation and double-diffusion processes.* Instability and breaking of internal waves trigger mixing and double-diffusion processes causing redistributions of temperature, salinity, and density in the upper ocean layer (Federov 1978). These circulations create the so-called nonequilibrium energy active zones favorable for the development of (sub)surface intrusions or temperature–salinity anomalies. Thermohaline fine processes are potentially unstable; as mentioned above, their surface manifestations can be detected by sensitive S–L band microwave radiometers.

Our theoretical analysis of experimental two-dimensional MSIW is based on a numerical simulation of passive microwave images. The idea is to generate digitally a field of the ocean brightness temperature $T_B(\vec{r})$ by a spatial field of surface currents $U(\vec{r})$ induced by internal waves. An analogical approach has been considered in microwave models operated with the wave action balance equation (Godin and Irisov 2003; Irisov 2007).

The implementation framework consists of three main parts: (1) hydrodynamic—the generation of disturbances by the surface current wave number spectrum; (2) electromagnetic—the computation of the brightness–temperature contrast; and (3) utilization of imaging data, including the comparison between model and experimental two-dimensional MSIW. Such a framework has also provided an observability estimate for the case of environmental internal wave events.

In the hydrodynamic part, a simplified solution called the "beta dominant approximation" of a wave action balance equation is used (Alpers and Hennings 1984; Liu 1988; Thompson et al. 1988). The following input parameters of the internal wave scene are considered: (U) is a set of individual steady solitons, (f) is a linearization perturbation spectral function, and (S) is the Phillips wave number spectrum of the surface:

$$U(\vec{r}) = \sum_{i=1}^{N} U_i(\vec{r}), \quad U_i(\vec{r}) = U_{0i} \operatorname{sech}^2\left(\vec{K}_i \vec{r} - \Psi_i\right), \tag{5.13}$$

$$f(\vec{k};x,y) = \frac{S(\vec{k};x,y) - S_0(\vec{k})}{S_0(\vec{k})} \approx -\gamma \frac{\partial U}{\partial x}, \tag{5.14}$$

$$S(\vec{k};x,y) = A(x,y)|\vec{k}|^{-n} Q(\varphi - \varphi_0), \tag{5.15}$$

where $U_i(\vec{r})$ is the current velocity induced by the *i*th soliton with wave number vector $\vec{K}_i$, peak U_{0i}, and phase Ψ_i; N is the number of participating solitons; $S(\vec{k};x,y)$ and $S_0(\vec{k})$ are perturbed and unperturbed wave number spectra, respectively; A, A_0, and n are parameters of the spectrum; $Q(\phi - \phi_0)$ is the spreading function of the spectrum, where ϕ and ϕ_0 are azimuthal angles; and $\gamma \approx 4.5/\beta$ is a constant, where β is the wind relaxation rate.

An electromagnetic part is based on the use of radiometric resonance model describing microwave emission from ocean-like rough surface (Irisov 1997, 2000, Section 3.32). The impact of internal wave events on surface microwave emission can be estimated as (Raizer 2007)

$$\Delta T_B(x,y) = 2T_0 k_0^2 A_0 \left(1 - \gamma \frac{\partial U}{\partial x}\right) B(k_0), \tag{5.16}$$

that is, the brightness–temperature contrast is proportional to the gradient of the surface current; $B(k_0)$ is a constant depending on the electromagnetic wave number k_0.

Now, the resulting discrete spectral image can be represented through an observation process (5.4) by a linear operator

$$T_I(i,j) \approx \mu G_U(i,j) + \eta, \tag{5.17}$$

where μ, η are random variables and $G_U(i,j)$ is a discrete gradient of $U(i,j)$. The set of numerical realizations—discrete radio-brightness scenes $T_I(i,j)$—is generated for each microwave frequency separately by input field $U(i,j)$ using standard procedures of gridding and digital interpolation. This particular image modeling and simulations are based on a *statistical characterization* of discrete radio-brightness field $T_I(i,j)$. Although an outlined technique seems to be simple enough, computations of MSIW are not trivial because there are some uncertainties in the choice of optimal pixel discretization of

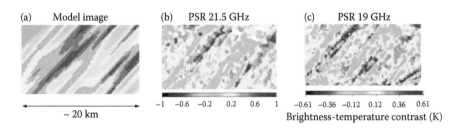

(a)   Model image   (b)   PSR 21.5 GHz   (c)   PSR 19 GHz

~ 20 km

-1   -0.6 -0.2   0.2   0.6   1   -0.61 -0.36 -0.12 0.12 0.36   0.61
                                         Brightness-temperature contrast (K)

FIGURE 5.25
Comparison between model and experimental data. Enhanced image fragments: (a) simulated digitally and (b) and (c) experimental (selected from PSR imagery, Figure 5.24). Radiometric signatures of are revealed in form of bright red stripes.

the scene (i.e., corresponding scale and grid resolution). The technique (5.17) yields an image change depending on the input parameters of the scene that can be used for numerical modeling and simulations of MSIW.

Figure 5.25 shows the comparison of (a) model and (b) and (c) experimental enhanced image fragments. In this particular example, a "multisoliton" one-dimensional model U(x; y = const) at large numbers N in sum (5.13) is used. The characteristics of internal waves are chosen in accordance with JUSREX'92 data (Gasparovic et al. 1993); parameters of the wave number spectrum are $n = 4$ and $A_0 = 10^{-3}$; $\beta = 2–3$.

As a result, periodic-like modulations of radio-brightness are generated perfectly that allows us to compare simulated and experimental MSIW. Calculated brightness-temperature contrasts therewith are $\Delta T_B = 1–2$ K or less as shown in color bars. It can be seen that an extremely good fit is achieved. More detailed interpretation requires involving more complicated hydrodynamic models.

A presented above, data demonstrates potential capabilities of high-resolution passive microwave radiometric technique for observing ocean internal waves. Focusing on the test PSR experiments conducted in the New York Bight in 2001, a geometrical similarity of radiometric and radar signatures of ocean internal waves has been evidenced. It is also possible to realize a link between model and experimental MSIW using a combined physics-based and digital image modeling. We believe, therefore, that combined active/passive microwave imaging techniques have the potential for the detection of deep-ocean wave phenomena.

5.5.3 Ship Wake Patterns

Ship detection is an important part of remote sensing applications needed for vessel traffic services, naval operations, and maritime surveillance. We all know that high-quality aerial photography and digital video provide incredibly detailed visualization of the ship wakes. Surface ship wakes are perfectly observed in SAR images (Alpers et al. 1981; Lyden et al. 1988; Eldhuset 1996; Stapleton 1997; Hennings et al. 1999; Fingas et al. 2001; Tunaley 2004; Soloviev et al. 2010; Brush et al. 2011). The configuration of the wake signatures in the radar images depends on environmental conditions significantly. The best results are obtained in the cases of calm and moderate winds.

Although the theory and detail description of ship wake phenomena are beyond the scope of this book; however, possible types of surface wakes can be pointed out. There are the following categories of sea surface wake patterns that are potentially detectable: (1) narrow V-wake, (2) classical Kelvin wake, (3) ship-generated internal wave wake, and (4) turbulent and vortex wake.

Ship wake can also be observed by passive microwave radiometers from shipborne platform or low-flying aircraft or helicopter. However, the characteristics of electromagnetic signals registered by microwave

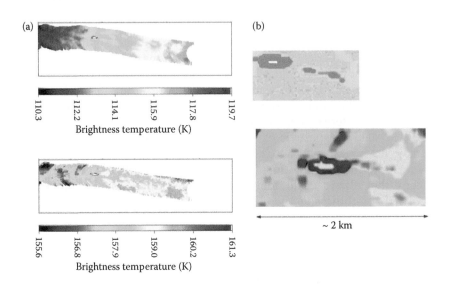

FIGURE 5.26

Microwave radiometric portrait of ship wake. (a) PSR images obtained from flight altitude 1500 ft. (b) Enhanced image fragment. Microwave signatures are blue spotted stripe and yellow cone.

radiometer and radar are different. Radar signatures are defined by Bragg-scattering effects, their specific geometrical modulations due to the excitation of surface waves from a moving ship. Radiometric signatures are defined by a number of factors: wake-generated surface roughness, turbulence, breaking waves, and foam/whitecap patterns. Therefore, radar and radiometric signatures of the ship wake have different structures and contrasts.

Figure 5.26 shows experimental realizations of passive microwave (PSR) observations of surface ship wake. These data were obtained from aircraft at low-altitude flight. Radiometric signatures of the wake represent narrow stripes located behind the ship. At the same time, it seems to be a V-type or Kelvin-type wake pattern of variable radio-brightness. These very first data allow us to assume that passive microwave radiometer is able to detect turbulent wake induced by the ship propeller and located very closely behind. The most contrast signatures are associated with the generated wave breaking and foam/whitecap patterns.

The structure of the wake is schematically shown in Figure 5.27. Two zones are distinguished: so-called near-wake and far-wake. The near-wake is well observed by radiometer and the far-wake is usually well observed by radar. Thus, again, a combined active/passive microwave technique is capable of providing a more robust detection and recognition of ship wakes. It is important especially for the monitoring of multiple-type vessels at coastal and harbor areas where the observed radar signatures may have unrecognizable configurations.

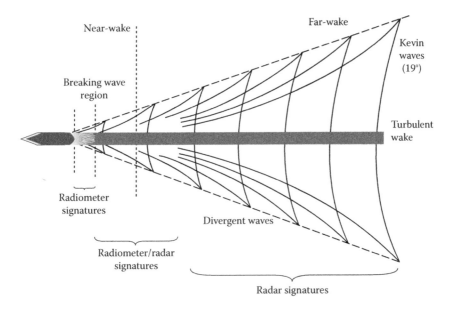

FIGURE 5.27
Schematic illustration of ship wake and radar/radiometer signatures. (Adapted and updated from George, S. G. and Tatnall, A. R. L. 2012. Measurement of turbulence in the oceanic mixed layer using Synthetic Aperture Radar (SAR). *Ocean Science Discussions*, 9:2851–2883. http://www.ocean-sci-discuss.net/9/2851/2012/osd-9-2851-2012-print.pdf.)

5.6 Summary

The goal of this chapter was to demonstrate benefits and advantages of high-resolution multiband passive microwave imagery. This technique yields microwave radiometric portrait (or picture) of the ocean surface. In the regular case of the absence of induced events, such a microwave picture reflects environmental conditions associated with large-scale ocean–atmosphere interactions and wind–wave dynamics. We define this situation as "ocean microwave stochastic background." Usually, an image background represents spotted mosaic textures of complicated geometry and variable radio-brightness. Multiband observations provide more objective and precise geophysical information. The best application of background data is still the retrieval of wind vector fluctuations in the restricted ocean areas.

The occurrence of the spotted microwave image textures in the presence of localized events can be explained by strong amplitude–frequency variations (modulations) of the surface wave number spectrum. In the case of internal waves, periodic-like radiometric signatures appear in the images due to spatial modulations of the roughness by surface current. In the case of ship wake, both effects of wave turbulence, wave breaking, and foaming result in the occurrence of multicontrast stripe-type and/or V-type radiometric signatures.

In the case of the presence of weakly visible or hidden (sub)surface events, more complicated analysis is required. First of all, the quality of microwave pictures should be improved significantly. This can be made using image enhancement algorithms. Second, it is necessary to develop and apply an interactive robust digital tool for the selection and extraction of relevant radiometric signatures from a stochastic image background. Such a tool includes statistical, correlation, fractal-based, and morphological operations. Computer vision algorithms can be involved as well (eventually, we deal with a computer vision product). Third, geophysical interpretation should be based on a combined theoretical–experimental approach involving modeling, simulation, and experimental verification of ocean microwave data. Statistical analysis and texture matching are the best options to provide the correct analysis. A theoretical–experimental approach (we name this thematic processing) allows us also to investigate properties of relevant signatures, develop their classifications, and create a *signature databank*. On this basis, the decision-making criteria can be established and employed for further needs.

The material represented in this chapter has demonstrated the potential ability of *multiband* passive microwave technique to observe localized and nonstationary ocean surface phenomena. The methodology is based on high-resolution, highly accurate geolocated registration of ocean microwave radiance, reliable acquisition of experimental data, their thematic processing, and digital evaluation of the signatures of interest. This research program requires some level of efforts.

References

Alpers, W. and Brümmer, B. 1994. Atmospheric boundary layer rolls observed by the synthetic aperture radar aboard the ERS-1 satellite. *Journal of Geophysical Research*, 99(C6):12613–12621.

Alpers, W. and Hennings, I. 1984. A theory of the imaging mechanism of underwater bottom topography by real and synthetic aperture radar. *Journal of Geophysical Research*, 89(C6):10529–10546.

Alpers, W. R., Ross, D. B., and Rufenach, C. L. 1981. On the detectability of ocean surface waves by real and synthetic aperture radar. *Journal of Geophysical Research*, 86(C7):6481–6498.

Apel, J. R. and Gonzalez, F. I. 1983. Nonlinear features of internal waves off Baja California as observed from the Seasat imaging radar. *Journal of Geophysical Research*, 88(7):4459–4466.

Basovich, A. Ya. and Tsimring, L. Sh. 1984. Internal waves in a horizontally inhomogeneous flow. *Journal of Fluid Mechanics*, 142:233–249.

Baum, E. and Irisov, V. 2000. Modulation of microwave radiance by internal waves: Critical point modeling of ocean observations. *IEEE Transactions on Geoscience and Remote Sensing*, 38(6):2455–2464.

Berizzi, F., Bertini, G., Martorella, M., and Bertacca, M. 2006. Two-dimensional variation algorithm for fractal analysis of sea SAR images. *IEEE Transactions on Geoscience and Remote Sensing*, 44(9):2361–2373.

Berizzi, F., Mese, E. D., and Martorella, M. 2004. A sea surface fractal model for ocean remote sensing. *International Journal of Remote Sensing*, 25(78):1265–1270.

Bettenhausen, M. H., Craig, K., Smith, G. K., Bevilacqua, R. M., Wang, N.-Yu., Gaiser, P. W., and Cox, S. 2006. A nonlinear optimization algorithm for WindSat wind vector retrievals. *IEEE Transactions on Geoscience and Remote Sensing*, 44(3):597–610.

Bindlish, R., Jackson, T. J., Gasiewski, A., Stankov, B., Klein, M., Cosh, M. H., Mladenova, I. et al. 2008. Aircraft based soil moisture retrievals under mixed vegetation and topographic conditions. *Remote Sensing of Environment*, 112(2):375–390.

Brush, S., Lehner, S., Fritz, T., and Soccorsi, M. 2011. Ship surveillance with TerraSAR-X. *IEEE Transactions on Geoscience and Remote Sensing*, 49(3):1092–1103.

Bulatov, M. G., Bolotnikova, G. A., Etkin, V. S., Skortsov, E. I., and Trokhimovsky, Yu. G. 1994. Shipborne microwave radiometer and scatterometer measurements of sea surface patterns during Joint US/Russia Remote Sensing Experiment. In *Proceedings of International Geoscience and Remote Sensing Symposium*, August 8–12, 1994, Pasadena, CA, Vol. 2, pp. 756–758.

Bunkin, F. V. and Volyak, K. I. 1987. *Oceanic Remote Sensing*. Nova Science Publisher Inc. (translated from Russian), Commack, New York.

Cavalieri, D. J., Markus, T., Hall, D. K., Gasiewski, A., Klein, M., and Ivanoff, A. 2006. Assessment of EOS Aqua AMSR-E Arctic sea ice concentrations using Landsat 7 and airborne microwave imagery. *IEEE Transactions on Geoscience and Remote Sensing*, 44(11):3057–3069.

Chapman, R. D. and Rowe, C. W. 1992. *Joint US/Russia internal wave remote sensing experiment. Meteorological data summary*. JHU/APL, Report SIR-92U-049. The Johns Hopkins University Applied Physics Laboratory, MD.

Cherny, I. V., Mitnik, L. M., Mitnik, M. L., Uspensky, A. B., and Streltsov, A. M. 2010. On-orbit calibration of the "Meteor-M" microwave imager/sounder. In *Proceedings of International Geoscience and Remote Sensing Symposium*, July 25–30, 2010, Honolulu, HI. pp. 558–561. Doi: 10.1109/IGARSS.2010.5651139.

Cherny, I. V. and Raizer, V. Y. 1998. *Passive Microwave Remote Sensing of Oceans*. Wiley, Chichester, UK.

Colliander, A., Lahtinen, J., Tauriainen, S., Pihlflyckt, J., Lemmetyinen, J., and Hallikainen, M. T. 2007. Sensitivity of airborne 36.5-GHz polarimetric radiometer's wind-speed measurement to incidence angle. *IEEE Transactions on Geoscience and Remote Sensing*, 45(7):2122–2129.

Dzura, M. S., Etkin, V. S., Khrupin, A. S., Pospelov, M. N., and Raev, M. D. 1992. Radiometers-polarimeters: Principles of design and applications for sea surface microwave emission polarimetry. In *Proceedings of International Geoscience and Remote Sensing Symposium*, May 26–29, 1992, Houston, TX, Vol. 2, pp. 1432–1434. Doi: 10.1109/IGARSS.1992.578475.

Eldhuset, K. 1996. An automatic ship and ship wake detection system for spaceborne SAR images in coastal regions. *IEEE Transactions on Geoscience and Remote Sensing*, 34(4):1010–1019.

Engler, O. and Randle, V. 2009. *Introduction to Texture Analysis: Macrotexture, Microtexture, and Orientation Mapping*, 2nd edition. CRC Press, Boca Raton, FL.

Ermakov, S. A., da Silva, J. C. B., and Robinson, I. S. 1998. Role of surface films in ERS SAR signatures of internal waves on the shelf. 2. Internal tidal waves. *Journal of Geophysical Research*, 103(C4):8033–8044.

Etkin, V. S., Trokhimovski, Yu. G., Yakovlev, V. V., and Gasparovic, R. F. 1994. Comparison analysis of Ku-band SLAR sea surface images at VV and HH polarizations obtained during the Joint US/Russia Internal Wave Remote Sensing Experiment. In *Proceedings of International Geoscience and Remote Sensing Symposium*, August 8–12, 1994, Pasadena, CA, Vol. 2, pp. 744–746. Doi: 10.1109/IGARSS.1994.399247.

Falconer, K. 1990. *Fractal Geometry: Mathematical Foundations and Applications*. John Wiley & Sons, UK.

Federov, K. N. 1978. *The Thermohaline Finestructure of the Ocean*. Pergamon Press, Oxford, UK.

Fingas, M. F. and Brown, C. E. 2001. Review of ship detection from airborne platforms. *Canadian Journal of Remote Sensing*, 27(4):379–385.

Franceschetti, G. and Riccio, D. 2007. *Scattering, Natural Surfaces, and Fractals*. Elsevier Academic Press, San Diego, CA.

Gade, M., Alpers, W., Wismann, V., Hühnerfuss, H., and Lange P. A. 1998. Wind wave tank measurements of wave damping and radar cross sections in the presence of monomolecular surface films. *Journal of Geophysical Research*, 103(C2):3167–3178.

Gaiser, P. W., Germain, K. M. St., Twarog, E. M., Poe, G. A., Purdy, W., Richardson, D., Grossman, W. et al. 2004. The WindSat spaceborne polarimetric microwave radiometer: Sensor description and early orbit performance. *IEEE Transactions on Geoscience and Remote Sensing*, 42(11):2347–2361.

Gasiewski, A. J., Piepmeier, J. P., McIntosh, R. E., Swift, C. T., Carswell, J. R., Donnelly, W. J., Knapp, E. et al. 1997. Combined high-resolution active and passive imaging of ocean surface winds from aircraft. In *Proceedings of International Geoscience and Remote Sensing Symposium*, August 3–8, 1997, Singapore, Vol. 2, pp. 1001–1005. Doi: 10.1109/IGARSS.1997.615324.

Gasparovic, R. F., Apel, J. R., and Kasischke, E. S. 1988. An overview of the SAR internal wave signature experiment. *Journal of Geophysical Research*, 93(C10):12304–12316.

Gasparovic, R. F., Chapman, R. D., Monaldo, F. M., Porter, D. L., and Sterner, R. E. 1993. *Joint U.S./Russia Internal Wave Remote Sensing Experiment: Interim Results*. JHU/APL Report S1R-93U-011. The Johns Hopkins University Applied Physics Laboratory, MD.

Gasparovic, R. F. and Etkin, V. S. 1994. An overview of the joint US/Russia internal wave remote sensing experiment. In *Proceedings of International Geoscience and Remote Sensing Symposium*, August 8–12, 1994, Pasadena, CA, Vol. 2, pp. 741–743. Doi: 10.1109/IGARSS.1994.399246.

Glazman, R. 1988. Fractal properties of the sea surface manifested in microwave remote sensing signatures. In *Proceedings of International Geoscience and Remote Sensing Symposium*, September 13–16, 1988, Edinburgh, Scotland, Vol. 3, pp. 1623–1624. Doi: 10.1109/IGARSS.1988.569545.

Glazman, R. E. and Weichman, P. B. 1989. Statistical geometry of a small surface patch in a developed sea. *Journal of Geophysical Research*, 94(C4):4998–5010.

Godin, O. A. and Irisov, V. G. 2003. A perturbation model of radiometric manifestations of oceanic currents. *Radio Science*, 38(4):8070–8080.

Gonzalez, R. C. and Woods, R. E. 2008. *Digital Image Processing*, 3rd edition. Prentice Hall, Upper Saddle River, NJ.

Grankov, A. G. and Milshin, A. A. 2015. *Microwave Radiation of the Ocean-Atmosphere. Boundary Heat and Dynamic Interaction*, 2nd edition. Springer, Cham, Switzerland.

Hennings, I., Romeiser, R., Alpers, W., and Viola, A. 1999. Radar imaging of Kelvin arms of ship wakes. *International Journal of Remote Sensing*, 20(13):2519–2543.

Irisov, V. G. 1997. Small-slope expansion for thermal and reflected radiation from a rough surface. *Waves in Random Media*, 7(1):1–10.

Irisov, V. G. 2000. Azimuthal variations of the microwave radiation from a slightly non-Gaussian sea surface. *Radio Science*, 35(1):65–82.

Irisov, V. G. 2007. Radiometric model of the sea surface in the presence of currents. *IEEE Transactions on Geoscience and Remote Sensing*, 45(7):2116–2121.

Ivanov, A. Y. and Ginzburg, A. I. 2002. Oceanic eddies in synthetic aperture radar images. In *Proceedings of the Indian Academy of Sciences-Earth and Planetary Sciences*, 111(3):281–295.

Jackson, C. R. and Apel, J. R. 2004. *An Atlas of Internal Solitary-Like Waves and Their Properties*, 2nd edition. Prepared under contract with Office of Naval Research. Code 322PO. Internet http://www.internalwaveatlas.com/Atlas_index.html

Jackson, T. J., Bindlish, R., Gasiewski, A. J., Stankov, B., Klein, M., Njoku, E. G., Bosch, D., Coleman, T., Laymon, C., and Starks, P. 2005. Polarimetric scanning radiometer C and X band microwave observations during SMEX03. *IEEE Transactions on Geoscience and Remote Sensing*, 43(11):2418–2430.

Kerman, B. R. and Bernier, L. 1994. Multifractal representation of breaking waves on the ocean surface. *Journal of Geophysical Research*, 99(C8):16179–16196.

Kerr, Y. H., Waldteufel, P., Wigneron, J.-P., Martinuzzi, J.-M., Font, J., and Berger, M. 2001. Soil moisture retrieval from space: The soil moisture and ocean salinity (SMOS) mission. *IEEE Transactions on Geoscience and Remote Sensing*, 39(8):1729–1735.

Klein, M., Gasiewski, A. J., Irisov, V., Leuskiy, V., and Yevgrafov, A. 2002. A wideband microwave airborne imaging system for hydrological studies. In *Proceedings of International Geoscience and Remote Sensing Symposium*, June 25–26, 2002, Toronto, Canada, Vol. 1, pp. 523–561.

Klemas, V. 2011. Remote sensing of sea surface salinity: An overview with case studies. *Journal of Coastal Research*, 27(5):830–838.

Klotz, B. W. and Uhlhorn, E. W. 2014. Improved stepped frequency microwave radiometer tropical cyclone surface winds in heavy precipitation. *Journal of Atmospheric and Oceanic Technology*, 31(11):2392–2408.

Kramer, H. J. 2002. *Observation of the Earth and Its Environment: Survey of Missions and Sensors*, 4th edition. Springer, Berlin.

Krasnopolsky, V. M., Breaker, L. C., and Gemmill, W. H. 1995. A neural network as a nonlinear transfer function model for retrieving surface wind speeds from the special sensor microwave imager. *Journal of Geophysical Research*. 100(C6):11033–11045.

Kropfli, R. A., Ostrovski, L. A., Stanton, T. P., Skirta, E. A., Keane, A. N., and Irisov, V. 1999. Relationships between strong internal waves in the coastal zone and their radar and radiometric signatures. *Journal of Geophysical Research*, 104(C2):3133–3148.

Kunkee, D. B. and Gasiewski, A. J. 1997. Simulation of passive microwave wind direction signatures over the ocean using an asymmetric-wave geometrical optics model. *Radio Science*, 32(1):59–77.

Kuzmin, A. V. and Pospelov, M. N. 1999. Measurements of sea surface temperature and wind vector by nadir airborne microwave instruments in Joint United States/Russia Internal Waves Remote Sensing Experiment JUSREX'92. *IEEE Transactions on Geoscience and Remote Sensing*, 37(4):1907–1915.

Le Vine, D. M. 1999. Synthetic aperture radiometer systems. *IEEE Transactions on Microwave Theory and Techniques*, 47(12):2228–2236.

Le Vine, D. M., Lagerloef, G. S. E., Coloma, R., Yueh, S., and Pellerano, F. 2007. Aquarius: An instrument to monitor sea surface salinity from space. *IEEE Transactions on Geoscience and Remote Sensing*, 45(7):2040–2050.

Le Vine, D. M., Lagerloef, G. S. E., and Torrusio, S. E. 2010. Aquarius and remote sensing of sea surface salinity from space. *Proceedings of the IEEE*, 98(5): 688–703.

Li, S. Z. 2001. *Markov Random Field Modeling in Image Analysis*. Springer, Tokyo.

Li, X., Clemente-Colón, P., Pichel, W. G., and Wachon, P. W. 2000. Atmospheric vortex streets on a RADARSAT SAR image. *Geophysical Research Letters*, 27(11):1655–1658.

Liu, A. K. 1988. Analysis of nonlinear internal waves in the New York Bight. *Journal of Geophysical Research*, 93(C10):12317–12329.

Lyden, J. D., Hammond, R. R., Lyzenga, D. R., and Schuchman, R. A. 1988. Synthetic aperture radar imaging of surface ship wakes. *Journal of Geophysical Research*, 93(C10):12293–12303.

Mallat, S. 2009. *A Wavelet Tour of Signal Processing: The Sparse Way*, 3rd edition. Academic Press, Burlington, MA.

Mandelbrot, B. B. 1983. *The Fractal Geometry of Nature*, 3rd edition. W. H. Freeman and Co, New York.

Martin, S. 2014. *An Introduction to Ocean Remote Sensing*, 2nd edition. Cambridge University Press, Cambridge, UK.

Mather, P. M. and Koch, P. 2011. *Computer Processing of Remotely-Sensed Images: An Introduction*, 4th edition. Wiley-Blackwell, UK.

Mirmehdi, M., Xie, X., and Suri, J. 2008. *Handbook of Texture Analysis*. Imperial College Press, London, UK.

Pesin, Y. and Climenhaga, V. 2009. *Lectures on Fractal Geometry and Dynamical Systems (Student Mathematical Library, volume 52)*. American Mathematical Society, Mathematics Advanced Study Semesters, Providence, RI.

Piepmeier, J. P. and Gasiewski, A. J. 1996. Polarimetric scanning radiometer for airborne microwave imaging studies. In *Proceedings of International Geoscience and Remote Sensing Symposium*, May 27–31, 1996, Lincoln, Nebraska, Vol. 3, pp. 1120–1122. Doi: 10.1109/IGARSS.1996.516587.

Piepmeier, J. P. and Gasiewski, A. J. 1997. High-Resolution multiband passive polarimetric observations of the ocean surface. In *Proceedings of International Geoscience and Remote Sensing Symposium*, August 03–08, 1997, Singapore, Vol. 2, pp. 1006–1008. Doi: 10.1109/IGARSS.1997.615325.

Piepmeier, J. P. and Gasiewski, A. J. 2001. High-resolution passive polarimetric microwave mapping of ocean surface wind vector fields. *IEEE Transactions on Geoscience and Remote Sensing*, 39(3):606–622.

Piepmeier, J. P., Gasiewski, A., Klein, M., Boehm, M., and Lum, R. 1998. Ocean surface wind direction measurements by scanning polarimetric microwave radiometry. In *Proceedings of International Geoscience and Remote Sensing Symposium*, July 6–10, 1998, Seattle, WA, Vol. 5, pp. 2307–2310. Doi: 10.1109/IGARSS.1998.702197.

Pratt, W. K. 2007. *Digital Image Processing*, 4th edition. John Wiley & Sons, Hoboken, NJ.

Porter, D. L. and Thompson, D. R. 1999. Continental shelf parameters inferred from SAR internal wave observations. *Journal of Atmospheric and Oceanic Technology*, 16(4):475–487.

Pospelov, M. N. 1996. Surface wind speed retrieval using passive microwave polarimetry: The dependence on atmospheric stability. *IEEE Transactions on Geoscience and Remote Sensing*, 34(5):1166–1171.

Qu, J. J., Gao, W., Kafatos, M., Murphy, R. E., and Salomonson, V. V. 2014. *Earth Science Satellite Remote Sensing: Vol. 1: Science and Instruments*. Springer-Verlag Berlin and Heidelberg GmbH & Co. KG.

Raizer, V. Y. 2001. Passive microwave radiometry, fractals, and dynamics. In *Proceedings of International Geoscience and Remote Sensing Symposium*, July 9–13, 2001, Sydney, Australia, Vol. 3, pp. 1240–1242. Doi: 10.1109/IGARSS.2001.976805.

Raizer, V. 2002. Statistical modeling for ocean microwave radiometric imagery. In *Proceedings of International Geoscience and Remote Sensing Symposium*, June 24–28, 2002, Toronto, Canada, Vol. 4, pp. 2144–2146. Doi: 10.1109/IGARSS.2002.10264721.

Raizer, V. 2003. Validation of two-dimensional microwave signatures. In *Proceedings of International Geoscience and Remote Sensing Symposium*, July 21–25, 2003, Toulouse, France, Vol. 4, pp. 2694–2696. Doi: 10.1109/IGARSS.2003.1294554.

Raizer, V. 2004. Correlation analysis of high-resolution ocean microwave radiometric images. In *Proceedings of International Geoscience and Remote Sensing Symposium*, September 20–24, 2004, Anchorage, Alaska, Vol. 3, pp. 1907–1910. Doi: 10.1109/IGARSS.2004.1370714.

Raizer, V. 2005a. Texture models for high-resolution ocean microwave imagery. In *Proceedings of International Geoscience and Remote Sensing Symposium*, July 25–29, 2005, Seoul, Korea, Vol. 1, pp. 268–271. Doi: 10.1109/IGARSS.2005.1526159.

Raizer, V. Y. 2005b. High-resolution passive microwave-imaging concept for ocean studies. In *Proceedings of MTS/IEEE OCEANS 2005 Conference*, September 18–23, 2005, Washington, D.C., Vol. 1, pp. 62–69. Doi: 10.1109/OCEANS.2005.1639738.

Raizer, V. 2007. Microwave radiometric signatures of ocean internal waves. In *Proceedings of International Geoscience and Remote Sensing Symposium*, July 23–27, 2007, Barcelona, Spain, pp. 890–893. Doi: 10.1109/IGARSS.2007.4422940.

Raizer, V. 2012. Fractal-based characterization of ocean microwave radiance. In *Proceedings of International Geoscience and Remote Sensing Symposium*, July 22–27, 2012, Munich, Germany, pp. 2794–2797. Doi: 10.1109/IGARSS.2012.6350852.

Raizer, V. Y. and Gasiewski, A. J. 2000. Observations of ocean surface disturbances using high-resolution passive microwave imaging. In *Proceedings of International Geoscience and Remote Sensing Symposium*, July 24–28, 2000, Honolulu, HI, Vol. 6, pp. 2748–2749. Doi: 10.1109/IGARSS.2000.859702.

Raizer, V. Y., Gasiewski, A. J., and Churnside, J. H. 1999. Texture-based description of microwave radiometric images. In *Proceedings of International Geoscience and Remote Sensing Symposium*, June 28–July 2, 1999, Hamburg, Germany, Vol. 4, pp. 2029–2031. Doi: 10.1109/IGARSS.1999.775022.

Rayzer, V. Yu. and Novikov, V. M. 1990. Fractal structure of breaking zones for surface waves in the ocean. *Izvestiya, Atmospheric and Oceanic Physics*, 26(6):491–494 (translated from Russian).

Raizer, V. Y., Novikov, B. M., and Bocharova, T. Y. 1994. The geometrical and fractal properties of visible radiances associated with breaking waves in the ocean. *Annales Geophysicae*, (12):1229–1233. Doi: 10.1007/s00585-994-1229-3.

Richards, J. A. and Jia, X. 2005. *Remote Sensing Digital Image Analysis: An Introduction*, 4th edition. Springer, Berlin.

Rosenfeld, A. and Lipkin, B. S. 1970. Texture synthesis. In *Picture Processing and Psychopictorics*. (Lipkin, B. C. and Rosenfeld, A., eds.), Academic Press, New York, pp. 309–345.

Ruf, C. S., Swift, C. T., Tanner, A. B., and Le Vine, D. M. 1988. Interferometric synthetic aperture microwave radiometry for the remote sensing of the Earth. *IEEE Transactions on Geoscience and Remote Sensing*, 26(5):597–611.

Russ, J. C. 1994. *Fractal Surfaces*, Plenum Press, New York, NY.

Russ, J. C. and Neal, F. B. 2015. *The Image Processing Handbook*, 7th edition. CRC Press, Boca Raton, FL.

Schertzer, D. and Lovejoy, S. 1991. *Non-Linear Variability in Geophysics: Scaling and Fractals*. Kluwer Academic Publishers, Dordrecht, the Netherlands.

Sharkov, E. A. 2007. *Breaking Ocean Waves: Geometry, Structure and Remote Sensing*. Praxis Publishing, Chichester, UK.

Shaw, J. A. and Churnside, J. H. 1997. Fractal laser glints from the ocean surface. *Journal of the Optical Society of America*, 14(5):1144–1150.

Shibata, A. 2006. A wind speed retrieval algorithm by combining 6 and 10 GHz data from Advanced Microwave Scanning Radiometer: Wind speed inside hurricanes. *Journal of Oceanography*, 62:351–359.

Shuchman, R. A., Lyzenga, D. R., Lake, B. M., Hughes, B. A., Gasparovic, R. F., and Kasischke, E. S. 1988. Comparison of Joint Canada-U.S. Ocean Wave Investigation Project synthetic aperture radar data with internal wave observations and modeling results. *Journal of Geophysical Research*, 93(C10):12304–12316.

Skou, N. and Le Vine, D. M. 2006. *Microwave Radiometer Systems: Design and Analysis*, 2nd edition. Artech House, Norwood, MA.

Smale, S. 1967. Differentiable dynamical systems. *Bulletin of the American Mathematical Society*, 73(6):747–817. Doi: 10.1090/S0002-9904-1967-11798-1.

Soloviev, A., Gilman, M., Young, K., Brusch, S., and Lehner, S. 2010. Sonar measurements in ship wakes simultaneous with TerraSAR-X overpasses. *IEEE Transactions on Geoscience and Remote Sensing*, 48(2):841–851.

Stankov, B. B., Cline, D. W., Weber, B. L., Gasiewski, A. J., and Wick, G. A. 2008. High-resolution airborne polarimetric microwave imaging of snow cover during the NASA Cold Land Processes Experiment. *IEEE Transactions on Geoscience and Remote Sensing*, 46(11):3672–3693.

Stapleton, N. R. 1997. Ship wakes in radar imagery. *International Journal of Remote Sensing*, 18(6):1381–1386.

Thompson, D. R., Gotwols, B. L., and Sterner II, R. E. 1988. A comparison of measured surface wave spectral modulations with predictions from a wave-current interaction model. *Journal of Geophysical Research*, 93(C10):12339–12343.

Trokhimovski, Y. G. and Irisov, V. G. 2000. The analysis of wind exponents retrieved from microwave radar and radiometric measurements. *IEEE Transactions on Geoscience and Remote Sensing*, 38(1):470–479.

Tunaley, J. K. E. 2004. Algorithms for ship detection and tracking using satellite imagery. In *Proceedings of International Geoscience and Remote Sensing Symposium*, September 20–24, 2004, Anchorage, Alaska, Vol. 3, pp. 1804–1807.

Ulaby, F. T. and Long, D. G. 2013. *Microwave Radar and Radiometric Remote Sensing*. University of Michigan Press, Ann Arbor, Michigan.

Wentz, F. J. 1992. Measurement of oceanic wind vector using satellite microwave radiometers. *IEEE Transactions on Geoscience and Remote Sensing*, 30(5):960–972.

Yueh, S. H. 1997. Modeling of wind direction signals in polarimetric sea surface brightness temperature. *IEEE Transactions on Geoscience and Remote Sensing*, 35(6):1400–1418.

Yueh, S. H., Wilson, W. J., Dinardo, S. J., and Hsiao, S. V. 2006. Polarimetric microwave wind radiometer model function and retrieval testing for WindSat. *IEEE Transactions on Geoscience and Remote Sensing*, 44(3):584–596.

Yuen, H. C. and Lake, B. M. 1982. Nonlinear dynamics of deep-water gravity waves. In *Advances in Applied Mechanics*, Vol. 22, pp. 67–229. Academic Press, New York.

6

Applications for Advanced Studies

In this chapter, the possibilities to observe a number of the so-called weakly emergency events are discussed. Research-based assessment is considered with the goal to determine the signature performance. The proposed material may seem counterintuitive but it does provide scientific and technological breakthroughs in the field of ocean microwave remote sensing. We believe that our ideas and predictions will be useful in future developments.

6.1 Surface Disturbances and Instabilities

A common hydro-physical flowchart of deep-ocean microwave diagnostics is shown in Figure 6.1. This problem is associated with the interaction between natural wind-wave processes and disturbances induced by a certain deep-ocean source (earthquake or explosion). Natural factors can enhance or depress the specified dynamical processes, may also have a pronounced effect on the lifetime of induced perturbations, and reduce the "theoretically" predicted hydrodynamic effect.

On the other hand, the oscillating character of the surface can cause correlations or decorrelations between multiband radiometric signals that provides a possibility to detect the interaction process and reveal time-dependent microwave signatures. In some cases, an amplification mechanism results in the generation of strong surface disturbances at restricted ocean areas that can be perceived as "roughness anomaly." An environmental example is surface *suloy* (rip currents). The corresponding microwave signatures can be evaluated using a resonance model of microwave emission (Section 3.3.2).

Abstracting from the hydrodynamic aspects of the problem, let us suppose that there are internal wave perturbations that propagate in the turbulent medium with unstable characteristics. For example, favorable conditions can arise due to the propagation of nonlinear internal waves in stratified upper ocean with unstable thermocline. It can be expected that the field of internal waves truly map onto the ocean surface. Under the influence of wave–current interaction, the development of modulation instabilities can occur that causes the excitation of high-frequency harmonics in the wave

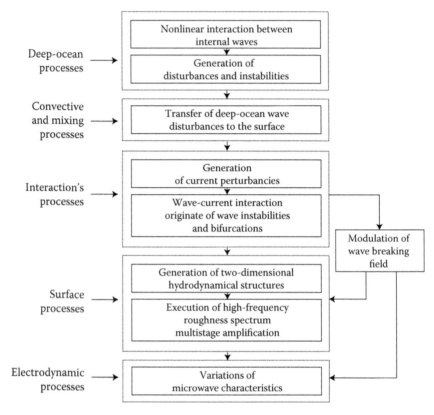

FIGURE 6.1
A hydro-physical concept of deep-ocean microwave diagnostics.

spectrum. If a cascade process takes place, it gives rise to the amplification of surface waves or the occurrence of roughness anomaly.

A scheme of the excitation process and possible microwave response is shown in Figure 6.2. The time of the existence of instabilities may be shorter than the time of active wave–current interaction, but the frequency of its origination may be high. The effects of stochastic autogeneration of "burst"-type surface disturbances are quite possible as well. In this case, we may observe an enhanced microwave response at selected microwave frequencies.

At a large time cumulation and spatial averaging of microwave signals, the effect of modulation instabilities will form the continuous-type microwave image with monotone radio-brightness characteristics. The appearance of multicontrast distinct regions (spots) in the image may correspond to localized surface roughness anomaly. Therefore, spot-type microwave signatures can also be indicators of certain deep-ocean processes. Similar microwave pictures have been observed by the PSR (Chapter 5).

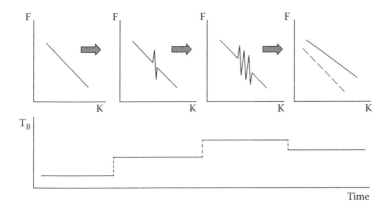

FIGURE 6.2
Diagram of cascade-type execution of wave number spectrum F(K) and corresponding response of microwave signal $T_B(t)$.

6.2 (Sub)Surface Wakes

This possibility is based on the passive microwave registration of a specific surface turbulent pattern, which is also known as "turbulent flow" or "turbulent wake." In most environmental situations, ocean turbulent wake represents a stochastic compact system of small-scale (horseshoe) vortexes with highly spatial variability.

The following effects can be possible indicators of turbulent wake: (1) generation of multimode surface wave spectrum at a localized ocean area due to interactions between turbulence and wind waves; (2) change of physical properties of the near-surface upper ocean layer under the influence of turbulent mixing processes, that is, the occurrence of the so-called mixed subsurface environment; (3) generation of Kelvin-type surface wakes; (4) generation of the Karman vortex street forming behind moving body; and (5) appearance of two-phase turbulent flows due to intensive cavitation or bubble activity. The last event is named "turbulent bubble wake." Turbulent wake (or collapsing turbulent wake) can also appear under the influence of a *distant source* of internal waves and their breaking.

We may expect to observe a variety of multicontrast radio-brightness signatures in high-resolution microwave images. The wake can be supposedly detected and recognized in the image by specific geometrical signatures, for example, in the form of nonlinearity, spotted, broken line, or narrow low-contrast stripe.

On the other hand, the interaction of turbulent wake with wind-generated surface waves may result in the signature transformation and/or its disappearing over time. As we all know, environmental conditions (wind action, atmospheric stratification, currents, and other natural factors) affect the dynamics of turbulent wake significantly. Therefore, reliable detection requires real-time observations.

6.3 Wave–Wave Interactions

Parametric interactions between surface gravity waves yield considerable variations of the wave spectrum. Resonance nonlinear interactions cause the generation of side-frequency wave components and the change of slope statistics of surface gravity and short gravity waves. These factors have an impact on the variations of the sea surface backscatter and emission.

Possible microwave signatures related to wave–wave interactions may represent grating-type radio-brightness textures in the form of geometrical sets of distinct short lines distributed at localized image regions. For a better detection performance, standard Fourier and correlation analyses can be applied.

Wave–wave interactions can also be observed using microwave radiometer-scatterometer. The signatures are defined using cross-spectrum analysis of the brightness temperature and the backscattering coefficient. A more complicated method is known as "mixed space–time spectral analysis"; it can be applied for the extraction of spectral signatures at the wave number–frequency domain.

As a whole, low-contrast spectral resonance-type signatures are associated with spatial hydrodynamic modulations, parametric wave interactions, or other surface excitation processes. Their reliable detection is possible using high-performance digital observation technology. Electro-optical sensors are capable of providing the type of information needed.

6.4 Thermohaline Anomaly

Thermohaline anomaly is a result of strong simultaneous variations (fluctuations) of sea surface temperature and salinity at localized ocean areas. Subsurface thermohaline fluctuations can occur under the influence of deep-ocean processes: internal wave breaking, double-diffusive convection, turbulent mixing transition (in the form of turbulent spots or intrusions), or other events. Horizontal variations of thermohaline fine structure may also occur due to strong currents, atmospheric precipitation, tropical rain, or hurricane impact.

A particular interest to us is the volumetric layering, or a clumping of the thermal and saline profiles in subsurface (1–2 m of depth) upper ocean. This process can form the so-called thermohaline wake, which is a floating dynamic cell pattern of varying salinity, temperature, and density.

Microwave manifestations of thermohaline fine structure may have different forms. The possible signatures represent quasi-regular cell-type patterns that appear as a result of interactions of thermohaline and wave processes. More sophisticated structures (complex patterns) are generated in the presence of strong surface turbulence or surface current (induced, for example, by internal waves). In both cases, an increment of ocean emissivity exists. We believe that the best instrument for the detection of thermohaline wake is a sensitive high-resolution S–L band radiometer-imager.

6.5 Internal Waves

Ocean internal waves are perfectly observed in radar (SAR) images. Surface manifestations can also be detected using passive microwave radiometers (Section 5.5.2). Typical signatures represent periodic-like variations of the brightness temperature. In the high-resolution images, they are observed as a system of contrast parallel stripes. These signatures are produced mostly due to wave–current interactions in the field of internal waves.

It is possible to observe individual solitons as well as *vertically propagating internal wavepackets*. The ocean thermocline, however, prevents such a movement. This circumstance is important for the manifestation of internal waves produced by submarines. Because the behavior of internal waves in deep and shallow water is different, the corresponding signatures have dissimilar contrast and configuration. The spatial distribution of signatures may vary depending on environmental conditions and time frame. The detection of internal waves and solitons is still a difficult observation task for passive microwave radiometry and imagery.

6.6 Wave Breaking Patterns

Instabilities and bifurcations of wind-generated surface waves cause the breaking phenomena. As a result, the strong transformation of the wave number spectrum in a wide interval of spatial frequencies occurs. Under these conditions, "burst"-type effects and excitations of the spectral components induced by a source may be masked or smoothed.

However, there occur large-scale surface effects, associated with modulations of gravity waves, wave breaking and foam/whitecap activity. These processes are observable by microwave and optical sensors.

The following effects can be considered in this connection:

- Originate "secondary" modulational instabilities due to the interaction between short and long surface waves
- Change of slope statistics of surface gravity waves, development of non-Gaussian distributions
- Generation of two- and three-dimensional (coherent) wave breaking patterns
- Increase the intensity (frequency) of wave breaking actions
- Change of geometry and statistics of the wave breaking field
- Generation of quasi-deterministic foam/whitecaps patterns

Gravity wave instabilities and breaking patterns accompanied by the change of the wave slope statistics usually result in an increase of the radio-brightness contrast. These large-scale dynamic effects can be revealed and distinguished using multiband radiometric measurements.

The change of the three last factors—wave breaking frequency, geometry, and statistics—of foam/whitecap patterns is registered by the microwave radiometer in the form of impulse-type time series or monotonic trend.

Wave breaking events are also a possible indicator of internal ocean processes. It is a well-known fact that the intensity and frequency of wave breaking acts increase in the presence of internal waves even at low and moderate winds.

Figure 6.3 illustrates a variant of complex stochastic ocean microwave scene simulated digitally. This scene involves both geometrical and volume factors: surface disturbances, wave breaking patterns, and foam/whitecap objects. A spatial distribution of the brightness level corresponds to a random law. Black–white gradations in the image reflect a possible regularity of the surface. The black areas correspond to foam/whitecaps structures; the white areas correspond to surface roughness. Their randomization yields extended or distinct microwave signatures of whimsical geometry.

Unlike the optical image where surface waves and foam/whitecaps are identified visually one-to-one, the microwave image may not display these structures directly. Thus, area/shape metrics suggested for digital analysis of optical images (Section 2.5.3) should be extended in cases of microwave "fat" or "thin" image objects. These metrics may not fit optical data due to the difference in microwave and optical radiance mechanisms. Since the emissivity of foam/whitecap depends on their microstructure and geometry, microwave data may reflect different dynamical stages of entire wave

FIGURE 6.3
Complex ocean microwave scene (computer simulation example).

breaking processes. In other words, sophisticated multispectral microwave imagery is able to provide more detailed specifications of wave breaking processes and foam/whitecap (coverage) patterns than low-resolution optical imagery.

7

Summary

The main goal of this book has been to describe in more detail the principles of microwave diagnostics of ocean environments. The author hopes that this book gave the reader great insight into the problem. Indeed, the material selected and reported in this book demonstrates the potential capabilities of passive microwave techniques for the detection of complex hydrodynamic processes and events. The methodology is based on the precise high-resolution mapping, selection, and digital evaluation of microwave radiometric signatures. However, to achieve this goal, we have to invoke an extended remote sensing technology, including sophisticated instrumentation and digital processing. In this context, this book briefly explains how to make the impossible *possible* and simultaneously, it reveals numerous scientific advances in passive microwave remote sensing of the ocean.

Thermal microwave emission of the ocean is formed under the influence of many environmental factors; roughly speaking, they are well known as the following: dielectric dispersion in seawater; geometry of the surface—the geometrical factor (surface waves and roughness); and two-phase media (foam/whitecap/spray/bubbles)—the volume factor. These and other important hydrodynamic factors and processes were discussed in *more detail than ever before* in the corresponding remote sensing literature (Chapter 2).

The first (dielectric permittivity) factor provides a basic level of the ocean brightness temperature depending on electromagnetic wavelength, incidence, and polarization of the emitted radiance; the geometrical factor yields low-contrast brightness temperature variations (up to 3–5 K depending on observation conditions); and the volume factor produces strong changes and fluctuations in ocean microwave emission (~10–20 K in real situations) depending on the structural and statistical characteristics of disperse layers covering the surface (known as foam and whitecap area fractions). The contributions from the surface waves are calculated using a number of diffraction model approaches and approximations describing scattering and emission of electromagnetic waves from a rough random surface with small- or large-scale irregularities. Statistical properties, probability distributions, and correlation functions of the sea surface elevation are involved as well. The contributions from two-phase disperse media—foam and whitecap—are estimated using macroscopic, wave propagation, and radiative transfer models or their combinations. The total microwave impact from both geometrical and volume factors on ocean emissivity can be defined using composition multifactor (usually two- or three-factor) models or semiempirical

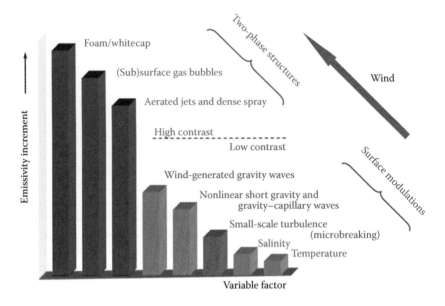

FIGURE 7.1
Summary diagram. Environmental factors contributed to ocean microwave emission.

approximations parameterized by wind speed. Figure 7.1 shows a summary diagram of the microwave contributions.

Geophysical interpretation involves regression estimation methods and multivariable techniques, depending on a given level of data representativeness. As a whole, the existing theoretical and data processing methods enable to provide adequate characterization of the *averaged* by space and time ocean emissivity at *selected* microwave frequencies and incidence angles (polarizations) and explain low-resolution (≥20–50 km) microwave radiometric observations (Chapter 3).

It appears that it is very easy to compute and/or evaluate the value of ocean emissivity using classical and well-known formulas; however, in real-world experiments conducted, for example, from the aircraft, we often observe a much more complicated microwave picture than it was drawn or predicted by the theory. This means that in addition to observation biases and instrument errors, there are other important "side factors" that contribute to emissivity in some way. These "hidden" microwave effects are not taken into account or simply neglected in most practical situations. This book outlines this problem very clearly in a scientific research manner. To do this, we refer, first of all, to the results of detailed numerical modeling (Chapter 4) and recent passive microwave observations (Chapter 5).

The entire high-resolution passive microwave portrait of the ocean surface looks complicated for quick analysis. A possible cause is *unpredictable intermittent noise* associated with environmental variability. Therefore, in order to extract the relevant information (the signatures of the interest) from such a stochastic hydrodynamic-microwave chaos, it is necessary to employ a special analytical tool. One option is a combined digital framework comprising both data processing and modeling (Chapter 4). By this means, advances of the interpretation are demonstrated as well.

We emphasize the importance of the following issues:

- High-resolution ocean microwave radiometric images are characterized by great texture variability. The main attributes are brightness mosaics and distinct spots.

- The occurrence of extended low-contrast image features is associated mostly with the ocean–atmosphere interaction, including dynamics of the ocean surface as well. Stochastic image mosaics represent a stochastic ocean microwave background.

- Deterministic image features are revealed in the form of distinct (hot and/or cold) spots of different shape, size, and brightness. Under certain conditions, they may represent microwave manifestations of localized ocean phenomena or events. For example, it could be surface roughness anomaly and/or wave breaking field.

- For adequate geophysical interpretation of microwave data, it is necessary to apply the combined methods of digital image analysis and computer modeling. Texture-fitting algorithms can provide an assessment of the signatures of interest.

- There are (de)correlations between multiband microwave images and signatures. These effects can be explained using multifactor models.

The common hydro-physics mechanism of ocean radio-brightness variations is strong amplitude–frequency transformations in the wave number spectrum (Chapter 6). For example, "burst"-type excitations and generation of side wave components at high-frequency spectral intervals may cause the formation of distinct radiometric signatures with correlated and/or decorrelated properties. Although direct measurement of spectral changes and intervals in the real world is highly improbable, computer modeling and simulations may reveal the main microwave effects.

Meanwhile, we found that standard methods of spectral and correlation analyses that are traditionally used in radar and optical studies, are also suitable for passive microwave radiometry and imagery. This fact was established a long time ago. However, a combined statistical correlation method allowed us to discover weak and strong correlations between extracted radiometric signatures, explore their multiband properties, and spatial distributions.

Another impressive digital technique considered in this book is based on the fusion of ocean multiband microwave data. We named this "ocean data fusion" (ODF). This well-known method was tested for ocean multiband (or potentially hyperspectral) microwave imagery for the first time. Although fusion is a more complicated procedure than correlation statistical analysis (simply because it is not obvious what radiometric channels should be fused to improve the signature performance), the technique demonstrates remarkable results. Our investigations show that parallel or hybrid data fusion network has significant advantages providing selection, enhancement, and/or elimination of informative data much better than the processing of selected one-channel data. For example, it is possible to extract radiometric signatures related to surface roughness anomaly or other given event and simultaneously reduce background effects. Therefore, we believe that the ODF is a promising and efficient tool for ocean remote sensing studies.

The scientific research and data reported in this book is just one step in the development of nonacoustic detection technology. Innovations in this field that are essential for passive/active microwave methods appear to be imperative.

Passive microwave remote sensing has the following advantages and benefits:

- No transmitted source
- Cannot be detected by active (radar) and other passive (infrared, video, optical) sensors
- All-weather day/night capability
- Penetration through Earth's atmosphere and cloudiness at low frequencies
- High sensitivity to sea state and hazard events
- Simultaneous monitoring of ocean and atmosphere parameters
- Sensitivity to (sub)surface mixing processes and two-phase flows
- Ability to provide their dielectric spectroscopy
- Multifrequency (sounding) polarimetric capability
- Wide swath and global coverage
- Instrument calibration stability
- Flexible low-power and low-mass technology
- Relatively low operational cost
- Simplicity of instrument modification and installation
- Long heritage and a variety of past applications

Objective disadvantages are difficulties to interpret and validate properly real-world ocean radiometric data and provide their overall application process. This book's chapters cover and discuss these issues extensively but not exhaustively.

A number of important practical and theoretical questions still remain. The main concern is the performance and optimization of high-resolution microwave measurements, utilization, specification, and assessment of the collected databases and signatures, as well as their thematic physics-based analyses. The reader may choose his/her own way to achieve a progress in this field.

Index

Printed and bound by CPI Group (UK) Ltd, Croydon, CR0 4YY

01/11/2024

01782619-0010